Vocational Education in the Fourth Industrial Revolution

James Avis

Vocational Education in the Fourth Industrial Revolution

Education and Employment in a Post-Work Age

palgrave
macmillan

James Avis
SEPD
University of Huddersfield
Huddersfield, UK

Institute of Education, University of Derby
Derby, UK

ISBN 978-3-030-52031-1 ISBN 978-3-030-52032-8 (eBook)
https://doi.org/10.1007/978-3-030-52032-8

This Palgrave Pivot imprint is published by the registered company Springer Nature Switzerland AG.
The registered company address is: Gewerbestrasse 11, 6330 Cham, Switzerland

Preface

The chapters in this book are written to be free-stranding. The book can be read in a linear fashion or readers may choose particular chapters that are of interest. Intentionally, the chapters are generously referenced, aiming to provide a resource directing readers attention to particular arguments they may wish to pursue further.

Debry, UK James Avis

Acknowledgements

The starting point for the book derives from my paper 'Socio-technical imaginary of the fourth industrial revolution and its implications for vocational education and training', published in the *Journal of Vocational Education & Training* (2018. 70:3, 337–363, DOI: 10.1080/13636820.2018.1498907).

I would like to acknowledge the support I have had from my family and partner, without which the book would be much poorer.

Contents

1

Introduction: Vocational Education in the Fourth Industrial Revolution: Education and Employment in a Post-Work Age

Abstract The introductory chapter provides an overview and signals the wide-ranging arguments the book addresses. It explores key terms, in particular Industrie 4.0 and the fourth Industrial Revolution (4th IR), examining the origins of the terms and their mobilisation. The 4th IR encompasses more than a narrow focus on advanced manufacturing; consequently this chapter includes discussions of post-work, Marxism and Italian Workerism which offer different understandings of the ideological and socio-economic context of the 4th IR to that found in policy and consultancy documents. Importantly, the chapter discusses vocational education and training and allied conceptualisations of skill. These arguments are related to the labour market and questions of class, gender and regional variations not only in Europe but also in low- and middle-income countries.

Keywords Fourth industrial revolution • Post-work • Ideology • Vocational education and training • Skill

J. Avis, *Vocational Education in the Fourth Industrial Revolution*,
https://doi.org/10.1007/978-3-030-52032-8_1

Introduction

Industrie 4.0 and the fourth Industrial Revolution (4th IR) anticipate the transformation of economic, social and political relations, or at least that is what some pundits would like us to believe. The terms themselves are fairly recent with Industrie 4.0 being coined in 2011 at the Hanover trade fair by three engineers,[1,2] and the 4th IR by Klaus Schwab of the World Economic Forum (WEF) in 2016 (Schulze 2019). In this book, I address a number of the debates and contestations that surround these terms. By doing so this allows me to consider the broader socio-economic context within which these debates are placed and the implications for the development of a socially just society. In addition, the chapter provides an overview of the other chapters, signalling the wide-ranging arguments that the book seeks to address.

Both terms Industrie 4.0 and the 4th IR find their way into any number of policy documents as well as in the pronouncements of consultancy firms, all of which call not only for the rethinking of organisational practices and the nature of waged labour but also of educational processes and vocational education and training (VET). More often than not the 4th IR is framed by a linear, technicist and deterministic logic in which the first Industrial Revolution is followed by the second, third and so forth. In this instance, diagrammatic representations similar to Fig. 1.1 are endlessly reiterated. These diagrams are reminiscent of those found in social and economic history textbooks and further secure the inevitability of these processes (Fuchs 2018; Peters 2017: 3; Kagermann et al. 2013: 13; Wahlster 2016: 4; Schrauf and Bertran 2016: 8). This trajectory is set within a very particular periodicy, with the first Industrial Revolution running from 1760 to around 1840 being related to the development of steam power, mechanical production and the railways. The second Industrial Revolution ran from the late nineteenth to the early mid-twentieth century and was marked by the development of mass production and electrification (Peters 2017: 3). The third Industrial Revolution featured the emergence of personal computers, digitalisation and the Internet and is seen as anticipating the 4th which intensifies these

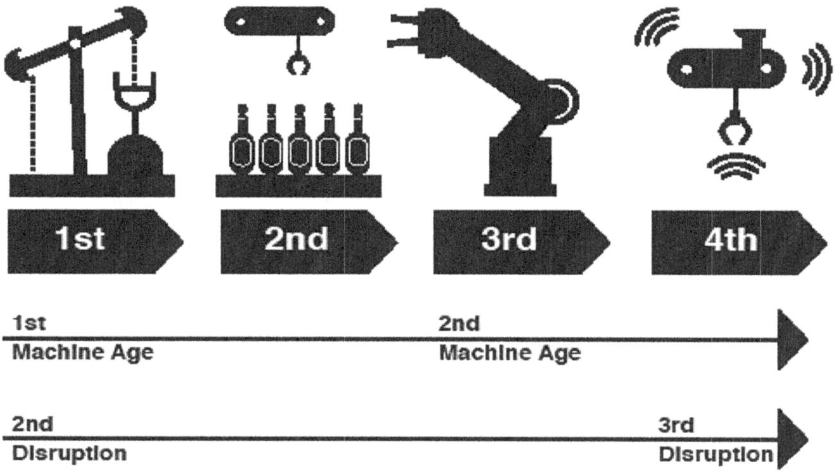

Fig. 1.1 Industrial revolutions

features. The 4th IR is construed as having a number of characteristics, which include (Schwab 2017: 7, 8)

> enabling "smart factories" … creating a world in which virtual and physical systems of manufacturing globally cooperate with each other in a flexible way. This enables the absolute customization of products and the creation of new operating models …
>
> It is the fusion of these technologies [gene sequencing, nanotechnology, quantum computing, etc.] and their interaction across the physical, digital and biological domains that make the fourth industrial revolution fundamentally different from previous revolutions.

Brynjolfsson and McAfee (2014) mobilise a not dissimilar trajectory arguing that the first Industrial Revolution ushered in the first machine age which was followed by the second machine age which developed in recent decades and is being taken forward in the current conjuncture. For these writers, the second machine age encompasses the 3rd and 4th IR and beyond. Bastani (2019) mobilises a rather different terminology and describes the current conjuncture as the third disruption, with the first representing the development of agriculture, the second the first Industrial

Revolution and the third disruption presaging a post-scarcity world (37). McAfee (2019) has argued, rather more conventionally, that the current conjuncture is one in which we can gain more from less, whereby fewer resources are required to produce socio-economic well-being.

There are a number of additional features associated with the 4th IR that are used to argue that we are entering a new era that is qualitatively different to the recent past. These features include the automation of not only manufacturing and industrial processes but also of those surrounding the provision of services and domestic life. These advances are allied to the development of smart factories and the Internet of Things (IoT). In this instance machines communicate with one another which in turn is related to the increasing salience of artificial intelligence, robotisation and the Internet of Services (IoS). The mobilisation of algorithms becomes all pervasive in any number of spheres from the factory, to finance, to dating web sites, to the home and so on. All of these ideas are linked to notions of interoperability, interconnectivity, automation and the development of networks (Hajkowicz et al. 2016; Bloem et al. 2014).

Pundits would argue we are on the cusp of a new era—the fourth Industrial revolution! The implication is that we should all adjust to the new conditions presaged by the 4th IR thereby ensuring our continuing viability as workers. Such a discourse is not only drawn upon in the global north but also the south in low- and middle-income countries where these are accompanied by calls for societal change. Whilst these new conditions are inevitably disruptive it is nevertheless suggested that they can be managed successfully to benefit all members of society. However, such a standpoint obscures the interests that lie behind such constructions, which are set on a very particular terrain. This terrain accepts unequivocally current social and economic relations, lodged within capitalism and a framework that is predicated on consensus which plays down the antagonistic relations between capital and labour, or in the vernacular, the conflictual relationship between the 1% and the rest—the 99%.

Socio-Economic Context

There are a number of socio-political and economic issue that serve to contextualise the 4th IR amongst which we could place neo-liberalism, globalisation and the transformation of waged relations, with the latter being echoed in the transition from Fordism to post-Fordism (Avis 2018). There are a number of narratives surrounding the development of neo-liberalism. In some accounts it developed as a response to the contradictions and failings of Keynesianism in the west and the social democratic welfare state which led to the fiscal crisis of the state in the 1970s, as well as growing rates of unemployment allied with the increasing costs of providing welfare services. In contradistinction to the universalism of social democracy with its concern with full employment and a fairer distribution of income and wealth, neoliberalism emphasises the 'free' market, the privatisation of public services, and the commodification of education and labour. In this context, lifelong learning is thought to be pivotal and can contribute to the ways we re-invent and invest in ourselves throughout our lifespan to meet the demands of an economically competitive society (Malloch et al. 2021). All of life is to be orientated towards the development of an economically competitive and efficient society. The claim being that such a social formation would be able to compete effectively in world markets, and if successful, all members of society would benefit. However, in the current conjuncture such aspirations have been shown to be deeply flawed with neo-liberalism exacerbating inequality leading to a society marked by an increasing polarisation between the rich and the rest of society.

In contrast to the determinism of the preceding linear description of the industrial revolutions (see Fig. 1.1) we could imagine another trajectory which runs from the depression of the pre-war years, through to the development of Keynesianism and social democracy following the Second World War and then on to neo-liberalism. However, in this instance the shift both to and from social democracy was not the result of technical determinism but rather derives from the balance of power between capital and labour. In other words, the outcome was forged through class struggle. The move towards neo-liberalism could be described in

Gramscian terms as transformism or as a passive revolution in which the dominant class transforms social relations so as to secure its interests. Yet in the current conjuncture neo-liberalism may be reaching its limits. We are, after Gramsci (1971: 276), entering an interregnum in which 'the old is dying and the new cannot be born' (and see Streeck 2016: 35–44). The flaws of neoliberalism are becoming increasingly apparent. To name but a few, these flaws can be seen in the failings of the market and of the privatisation of public services, the growth in inequality, the arbitrariness surrounding the distribution of income and wealth, the increasing polarisations in society, not only between rich and poor but also politically and socially. These flaws can also be seen in the retreat from the 'free' market allied with societal retrenchment whereby states become increasingly inward looking as best exemplified by Trump's America (Monbiot 2017). The current conjuncture is marked by the decoupling of employment and wages from productivity and growth, this is especially the case in Anglophone societies of the USA and UK and serves to undermine the claims of those who celebrate the operation of the market.

Neo-liberalism and its doctrine of market fundamentalism which fantasied about a minimal state, a meritocratic social order and a personally liberated world are seen as just that—a fantasy (Bloom and Sancino 2019: 19–22). For Mason (2019: xii) 'the free market system has imploded' and its mantra of consumerism, selfishness and hierarchy has lost its ideological and persuasive weight. Perhaps such accounts exaggerate the success of neo-liberalism which whilst all pervasive failed to secure the hegemony that social democracy had attained in the immediate post-war years in the west (Education Group 1981). Another way in which we can think about these processes and trajectories is through the alleged transition from industrial to immaterial capitalism which emphasises the salience of knowledge, digitalisation as well as the pivotal importance of cognitive and soft skills. In this instance, there is something of an affinity with the 4th IR as well as with the transition from Fordism to post-Fordism and beyond and is allied to discussions of the knowledge/information society, in which intellectual property rights assume greater importance. Here, rather than outcomes being determined by the presumption of linear and technicist developments these latter arguments recognise that they are shaped by economic, social and political processes.

Italian Workerism is one way of attempting to make sense of these processes whereby worker struggles against the Fordist regime of industrial capitalism—the refusal of work—resulted in the shift towards post-Fordism and immaterial capitalism (Lotringer and Marazzi 2007; Tronti 2019).

The salience of immaterial labour has an affinity with the 4th IR and debates concerned with the nature of waged labour in current conditions which emphasise knowledge, cognitive processes and soft skills. It is important to acknowledge the material basis of immaterial labour as it is very easy to ignore the material infrastructure that underpins the Internet and cognitive labour. In addition features of the 4th IR, big data, AI, automation point in a number of different directions, towards both technological unemployment as well as the need for an increasingly skilled labour force. In these instances such changes could be seen as presaging transformative changes and impacting upon the labour market. In the case of technological unemployment this could be seen as deriving from the automation of routine manufacturing and white-collar jobs and their articulation with the IoT and IoS. These changes have been construed by some writers as anticipating a post-work society or at least one in which the hours of waged labour are considerably shortened providing space for the fuller development of human potentiality. Such developments could be seen as being undermined by the current economic and social relations of capitalism that are predicated on scarcity rather than the abundance promised by post-work and post-capitalist societies and the technological advances predicated by the 4th IR (Mason 2015).

The preceding discussion could be seen as rooted in the west and global north. There are however a number of points to be made. Capitalism of whatever type is orientated towards the accumulation of capital. The decision to mobilise machines or waged labour will not only be shaped by the balance of power between capital and labour but also by the resulting costs, which in turn will be related to the operation of the labour market. Thus, in uncertain times it may be economically expedient for capital to mobilise waged labour even though it might be more productive and cheaper to use machines and the IoT. Such processes will impact upon low- and middle-income countries and their use as sources of cheap and semi-skilled labour. In addition, Brown et al. (2011: 64) refer to 'oasis operations' whereby high-tech factories, offices and research institutes exist alongside 'low-spec' neighbourhoods upon which they have little effect. Such operations may nevertheless impact upon employment and wages in the developed nations by creating a high skills low-wage nexus. In this way such developments impact upon labour markets that are differentiated both within and across nations, regions and localities. This can be seen in the aforementioned 'oasis operations' whereby a globalised labour market can be set in the same locality as one that addresses a more localised labour market for low waged, low skilled as well as craft-based work (Peri 2018).

The forms of technological development presaged by the 4th IR can effect low- and middle-income countries in several ways. Historically, the development of industrial manufacturing plants served to provide significant benefits for low- and middle-income societies through the provision of employment and wages. However, late industrialisers deindustrialise at an earlier point than their earlier counterparts, thereby truncating these benefits. In addition such technological developments may facilitate the re-bundling of processes that were formerly separated leading to the reshoring of manufacturing plants in the developed societies and consequently resulting in a loss of employment in the emerging economies and depriving them of the opportunity to develop large manufacturing sectors. There is of course an irony with this argument which reads a little like that of an apologist for neo-liberal capitalism as it belies the suffering and exploitation that surrounds factory work in the developing and emerging economies. For whilst the move to urban settings in search of employment can be an antidote to extreme rural poverty it carries its own significant emotional and 'economic' costs which could barely be conceived of as an escape from poverty (War on Want 2019a, b).

Vocational Education and Training, Knowledge and Skill

I use the term VET to refer to education and training that is thought to directly address the needs of employers. Various terms have been used to describe this type of education, sometimes it is referred to a TVET with T standing for technical or career education. Clarke and Winch (2007: 1) suggest TVET is concerned with 'the social development of labour' implicitly pointing towards the reproduction and production of labour power. In the English context, VET has frequently been thought of as narrowly and instrumentally orientated towards work and as having relatively low status, being best suited to 'other people's children'. Even when the term is used to consider university education and is described as higher vocational education (HIVE) it often echoes the preceding disparagement. Bathmaker (2014: np) suggests VET includes 'pre-vocational preparation, general vocational education, work-related learning, work-based learning, higher vocational education (for associated professionals and technicians)', distinguishing it from academic and professional education. However, if we consider HIVE and VET as being concerned with the social development of labour it operates on a much broader sphere and will encompass professional education as well as much that passes as higher education. It is the case, that in recent years and particularly, following the 2008 financial crisis that education in general and higher education in particular has become increasing vocationalised and is to become business facing which is now an important feature of the education landscape.

The rhetoric that surrounds VET, knowledge and skill in the current conjuncture mimics the types of arguments that were present in the last half of the twentieth century. Here we encounter a range of familiar themes that discuss the constant need for up- and reskilling with the resulting necessity that workers are flexible and should engage in lifelong learning. These notions sit alongside others that address the changing nature of waged employment and the shift away from manufacturing towards the service sector. Here again we meet with well-established neo-liberal tropes that call for the development of an enterprising subjectivity, which would, or so it is claimed, develop skills and dispositions that enable workers to survive the precarious labour market conditions they encounter. An emphasis is given to the requirement for the development not only of cognitive but also soft skills. There are two points to raise; first, the rapidity of technological change demands the acquisition of skills that enable workers to manage the ongoing transformations that impact upon waged labour. Second, whilst employment in manufacturing may be less widespread than in the past, it still requires the development of technological skills that would enable workers to surveil the IoT and to handle any crises that occur in manufacturing processes. This would require the development of high-level cognitive and problem-solving skills together with a specific and complex understanding of the particular manufacturing process and its configuration. Thus, the rapidity of technical changes requires the development of generic and easily transferable skills which sits somewhat contradictorily alongside the need for skills that address the specificity of any particular manufacturing process. These issues apply not only to high but also to medium- and low-income countries and will be accented according to the specific context in which they are located.

In many accounts concerned with the developing or emerging economies of the global south the rhetoric of up-skilling rest alongside the perception of a skills-gap. In this instance questions of diversity and equality are raised with women and girls being seen as an important reservoir of labour that could be mobilised and up-skilled thereby addressing the skills-gap. Such processes are uneven and encounter the very particular labour markets that are confronted, both formal and informal, which are set in specific socio-economic and political contexts. It is important to acknowledge that alleged skills-gaps and labour shortages can co-exist alongside the existence of surplus labour and it is important to note, there are political as well as economic processes at play that relate to capitalist accumulation practices. In addition and importantly, discussions of the 4th IR may lead us to focus on manufacturing processes and consequently ignore or at least play down the very real significance of various forms of care and service labour that are found in the economies of both the global north and south and that are crucial to societal and individual well-being. At the time of writing, April 2020, we are living through the health crisis spawned by COVID-19. This crisis has disrupted social and economic relations across the globe with many social formations breaking free from the tenets of neo-liberalism leading to the development of interventionist states that echo aspects of social democracy. Vast amounts of money are being poured into economies in an attempt to secure the long-term interests of capital and the quiescence of those who have been furloughed, rendered redundant or face unemployment. At the same time although there is a recognition of the importance of waged and particularly unwaged labour, VET remains lodged in a neo-liberal and technicist framework that implicitly accepts the logic of competitiveness allied to the development of human capital. As Anderson (2009) argued some years ago TVET is embroiled in a productivist paradigm predicated upon economic growth and waged work which silences the manner in which TVET could serve alternative conceptualisations of labour. As yet it is uncertain as to how the pandemic will play out and the manner in which the crisis will impact on social and economic relations, or indeed the development of VET. Chapter 5 is followed by a postscript which addresses these questions.

Chapter Break Down

Chapter 1—Introduction: Vocational Education in the Fourth Industrial Revolution: Education and Employment in a Post-Work Age

This chapter serves as an introduction to the book.

Chapter 2—Socio-Technical Imaginaries and the Fourth Industrial Revolution

Those who promulgate notions of the 4th IR are at one and the same time constructing and reconstituting a particular understanding of the future and importantly, its imagining. This is often an iterative and circular process with the same arguments being reproduced in a range of policy and consultancy documents. Consequently, we encounter a number of recurring themes, robotisation, artificial intelligence, digitalisation, smart factories, interoperability, interconnectivity, nanotechnology and so on. Such notions seep into popular culture and media representations and contribute towards socio-technical imaginaries. Importantly, the imaginaries that swirl around the 4th IR acknowledge a range of contradictory tensions that either intensify current trends or are thought will emerge in the future. The chapter examines arguments derived from the WEF, the EU and the ILO that address the future of work and skills and draws out the implications of these for VET. In addition, it explores how these arguments play out in low- and middle-income economies. However, within this socio-technical imaginary there is the promise that difficulties can be managed and in this sense overcome. Whilst the discourse emphasises the transformative power of technology it is in effect domesticated, its potential framed to sit within a particular, and one might suggest, neoliberal or at least capitalist project.

Chapter 3—Robotisation, Artificial Intelligence, Employment and the Fourth Industrial Revolution

Any number of papers and books have argued that the growth of robotisation will have a profound impact on employment and will lead to the replacement of routine manual jobs, resulting in technological unemployment. This is set against those who draw upon historical examples to suggest that such arguments are misfounded. These arguments are echoed by labour market analysts of western societies who anticipate frictional/temporary unemployment before stability is reasserted. This chapter critiques, challenges and moves beyond such arguments which are distanced from the lived experience of workers in disadvantaged regions of Europe and the US. In order to fully understand these processes it is necessary to engage with arguments that address the capitalist labour process, class struggle as well as the significance of rentier capitalism. These arguments need to be articulated with conceptualisations of class, race and gender and their intersectionality in current conditions. In addition, this discussion needs to be related to the socio-economic conditions faced by low- and middle-income economies. The chapter concludes by considering the implication of the preceding discussion for VET.

Chapter 4—Post-Work, Post-Capitalism and the Fourth Industrial Revolution

This chapter draws upon the arguments raised earlier that address the future of work and skill setting this alongside more radical analyses. Much of the work that addresses the labour market refuses the suggestion that labour is being expelled from paid employment and stands in direct opposition to some of the analyses considered in the previous chapters. There is a tension between those who experience and bemoan precariousness, intermittent and under-employment and those for whom the whole of life seems to be centred around work. Paradoxically, for some this had led to the expulsion from waged labour, whilst for others it has led to its intensification. In its ideological manifestation the 4th IR is presented as an inevitable development flowing from the first to the current IR. However, what is silenced in these constructions is the manner in which the 4th IR connects to the development of capitalism and its articulation with class struggle. These processes alert us to the manner in which technology is entwined with social relations. Consequently, this chapter addresses arguments that engage with Italian Workerism, post-capitalism, post-work, universal basic income and paradoxically, the politics of abundance. Thus, it seeks to develop a politically informed analysis that moves beyond current conceptualisations of the 4th IR that in turn informs a critical, if not transformative praxis. The chapter calls for an expansive understanding of both class and VET that goes beyond a concern with waged labour.

Chapter 5—Conclusion: Vocational Education in the Fourth Industrial Revolution—Education and Employment in a Post-Work Age

This chapter concludes the book and serves to bring together the various themes that run through the book drawing out their significance for VET and educational systems committed to the development of socially just societies. At the same time such interventions have to go beyond the narrowly educative and engage with a broader politics committed to the transformation of society. Many of the rhetorical claims surrounding the 4th IR align themselves with a notion of social justice. However, this is set within a capitalist terrain that seeks to manage the tensions flowing from the new condition presaged by the 4th IR, rather than their transcendence. Such rhetorical claims frequently serve to domesticate tensions thereby underplaying their potential in the struggle for the creation of a socially just society.

Postscript: COVID-19

This postscript examines a number of the debates about the health crisis and the impact of COVID-19 on social and economic relations as well as VET. Currently the tenets of neo-liberalism have been pushed to one side, replaced by the actions of an interventionist state seeking to secure the long-term interests of capital. Many commentators have argued that the pandemic will fundamentally transform social, political and economic relations. These stances vary between those adopting a pessimistic position whilst others have argued that the pandemic opens-up the possibility for the development of a kinder, more egalitarian socialist society. However, whatever the eventual outcome it will be shaped by the balance of power between labour and capital as well as by the progression of the pandemic.

Conclusion

In this introductory chapter, I have alluded to a number of the debates and questions that are to be addressed throughout this book and have included a postscript that considers the health crisis and COVID-19. The key concern being to critique constructions of the 4th IR through the lens of a socially just society. This leads to an examination of the manner in which we make sense of waged labour and its articulation to post-work and post-capitalism. This serves to raises the question as to whether we consider waged labour to be a site in which we enact our human potentiality or one where we experience exploitation and oppression. The association of waged work with job satisfaction and personal fulfilment could be seen as a way in which particular groups of workers are co-opted by capital. Such arguments bear on the manner in which VET, knowledge and skill is understood. Perhaps waged labour is so wedded to capitalist relations that in order to go past it we need to move beyond such labour towards a post-scarcity society—that is if we are committed to the development of a socially just society.

Acknowledgement This chapter draws in part on my 'Socio-technical imaginary of the fourth industrial revolution and its implications for vocational education and training: a literature review', published in the Journal of Vocational Education & Training, 2018. 70:3, 337–363, https://doi.org/10.1080/13636820.2018.1498907.

Notes

1. 'The marketing-style term Industrie 4.0 was invented and promoted by three engineers: Henning Kagermann (physicist and one of the founders of SAP), Wolfgang Wahlster (professor of artificial intelligence), and Wolf-Dieter Lukas (physicist and senior official at the German Federal Ministry of Education and Research). What they started in 2011 during a press conference at the Hannover Messe and in 2016 became the main motto of the WEF meeting in Davos almost perfectly matches what McCray [2012] defines as visioneering' (Pfeiffer 2017: 107–108).

2. 'Fourth Industrial Revolution or Industry 4.0 was first mentioned by Bosch at the Hannover Trade Fair in 2011. Experts at the trade fair said that a new Industrial Revolution had arrived with innovations brought into production by the modern face of the information era. When the German Government took these opinions seriously, the Fourth Industrial Revolution had become official. After the trade fair, a working group on the Fourth Industrial Revolution was established. One year later, this working group presented its suggestions for the actual implementation of Industry 4.0 at the next Hannover Trade Fair, and reported this to the German Government. Bosch executive Siegfried Dais and SAP AG executive Henning Kagermann co-chaired the working group' (Bosch Turkey: online). https://www.sanayidegelecek.com/en/sanayi-4-0/tarihsel-gelisim/.

References

Anderson, D. 2009. Productivism and Ecologism: Changing Dis/courses in TVET. In *Work, Learning and Sustainable Development*, ed. J. Fien, R. Maclean, and Man-Gon Park, 35–57. Dordrecht: Springer.

Avis, J. 2018. Socio-Technical Imaginary of the Fourth Industrial Revolution and its Implications for Vocational Education and Training: A Literature Review. *Journal of Vocational Education & Training* 70 (3): 337–363. https://doi.org/10.1080/13636820.2018.1498907.

Bastani, A. 2019. *Fully Automated Luxury Communism*. London: Verso.

Bathmaker, A.-M. 2014. *'Applied', 'Technical' and 'Vocational' Constructions of Knowledge in Vocational Education*. Paper presented at Vocational Education and Training: Policy, Pedagogy and Research, BERA Post-Compulsory and Lifelong Learning SIG, The University of Birmingham.

Bloem, J., M. Van Door, S. Divesting, D. Excoffier, R. Maas, and E. Van Ommeren. 2014. *The Fourth Industrial Revolution Things to Tighten the Link between IT and OT*. VINT Research Report 3. https://www.fr.sogeti.com/globalassets/global/downloads/reports/vint-research-3-the-fourth-industrial-revolution.

Bloom, P., and A. Sancino. 2019. *Disruptive Democracy*. London: Sage Swifts.

Brown, P., H. Lauder, and D. Ashton. 2011. *The Global Auction*. Oxford: Oxford University Press.

Brynjolfsson, E., and A. McAfee. 2014. *The Second Machine Age: Work, Progress and Prosperity in a Time of Brilliant Technologies.* New York: Norton & Company.

Clarke, L., and C. Winch. 2007. Introduction. In *Vocational Education*, ed. L. Clarke and C. Winch, 1–18. London: Routledge.

Education Group, CCCS. 1981. *Unpopular Education*. London: Hutchinson.

Fuchs, C. 2018. Industry 4.0: The Digital German Ideology. *tripleC* 16 (1): 280–289. https://doi.org/10.31269/triplec.v16i1.1010.

Gramsci, A. 1971. *Selections from the Prison Notebooks.* London: Lawrence and Wishart.

Hajkowicz, S., A. Reeson, L. Rudd, A. Bratanova, L. Hodgers, C. Mason, and N. Boughen. 2016. *Tomorrow's Digitally Enabled Workforce: Megatrends and Scenarios for Jobs and Employment in Australia Over the Coming Twenty Years.* https://www.acs.org.au/content/dam/acs/acs-documents/16-0026_DATA61_REPORT_TomorrowsDigiallyEnabledWorkforce_WEB_160128.pdf.

Kagermann, H., W. Wahlster, and J. Helbig. 2013. *Recommendations for Implementing the Strategic Initiative INDUSTRIE 4.0, Securing the future of German manufacturing industry Final report of the Industrie 4.0 Working Group*, http://www.acatech.de/fileadmin/user_upload/Baumstruktur_nach_Website/Acatech/root/de/Material_fuer_Sonderseiten/Industrie_4.0/Final_report__Industrie_4.0_accessible.pdf.

Lotringer, S., and C. Marazzi. 2007. *Autonomia: Post-Political Politics.* Los Angeles: Semiotext(e).

Malloch, M., L. Cairns, K. Evans, and B. O'Connor, eds. 2021. *The SAGE Handbook of Learning and Work.* London: Sage.

Mason, P. 2015. *Postcapitalism*. London: Allen lane.

———. 2019. *Clear Bright Future*. London: Allen lane.

McAfee, A. 2019. *More from Less*. London: Simon and Schuster.

McCray, P.W. 2012. *The Visioneers: How a Group of Elite Scientists Pursued Space Colonies, Nanotechnologies, and a Limitless Future.* Princeton: Princeton University Press.

Monbiot, G. 2017. *Out of the Wreckage*. London: Verso.

Peri, N. 2018. *Examining Non-formal Education in Complex Knowledge Production: A Case Study of the Safety Match work and workers in South India.* Phd Thesis, Faculty of Social Sciences and Law, University of Bristol.

Peters, M.A. 2017. Technological Unemployment: Educating for the Fourth Industrial Revolution. *Educational Philosophy and Theory* 49 (1): 1–6. https://doi.org/10.1080/00131857.2016.1177412.

Pfeiffer, S. 2017. The Vision of "Industrie 4.0" In the Making—A Case of Future Told, Tamed, and Traded. *Nanoethics* 11: 107–121. https://doi.org/10.1007/s11569-016-0280-3.

Schrauf, S., and P. Bertran. 2016. *Industry 4.0: How Digitization Makes the Supply Chain More Efficient, Agile, and Customer-Focused.* PWC. https://www.strategyand.pwc.com/reports/digitization-more-efficient.

Schulze, E. 2019. *Everything You Need to Know About the Fourth Industrial Revolution.* https://www.cnbc.com/2019/01/16/fourth-industrial-revolution-explained-davos-2019.html.

Schwab, K. 2017. *The Fourth Industrial Revolution.* London: Portfolio Penguin.

Streeck, W. 2016. *How Will Capitalism End?* London: Verso.

Tronti, M. (2019 [1966]). *Workers and Capital.* Verso: London.

Wahlster, W. 2016. *Industrie 4.0: Cyber-Physical Production Systems for Mass Customization.* German-Czech Workshop on Industrie4.0/Průmysl 4.0, Prague, April 11. https://www.google.com/search?q=Industrie4.0%3A+Cyber-Physical+Production+Systems+for+Mass+Customization&rlz=1C1GCEA_enGB842GB842&oq=Industrie4.0%3A+Cyber-physical+Production+Systems+for+Mass+Customization&aqs=chrome..69i57.1070j0j7&sourceid=chrome&ie=UTF-8.

War on Want. 2019a. *Sweatshops in China.* https://waronwant.org/sweatshops-china.

———. 2019b. *Sweatshops in Bangladesh.* https://waronwant.org/sweatshops-bangladesh.

2

Socio-Technical Imaginaries and the Fourth Industrial Revolution

Abstract Those promulgating the fourth Industrial Revolution are simultaneously constructing and reconstituting a particular understanding of the future. This is often an iterative process with the same arguments being reproduced in policy and consultancy documents. We encounter a number of recurring themes, robotisation, artificial intelligence, digitalisation, smart factories, interoperability, interconnectivity, nanotechnology and so on. These seep into popular culture, contributing towards socio-technical imaginaries that acknowledge a range of contradictory tensions that either intensify current trends or are thought will emerge in the future. The chapter examines arguments about the future of work as well as skills, drawing out the implications for vocational education and training. Although these discourses emphasise the transformative power of technology this is domesticated, its potential framed within a particular neo-liberal or at least capitalist project.

Keywords Fourth industrial revolution • Socio-technical imaginaries • Future of work and skill • Neo-liberalism • Capitalism

J. Avis, *Vocational Education in the Fourth Industrial Revolution*,
https://doi.org/10.1007/978-3-030-52032-8_2

Introduction

The preceding chapter examined conceptualisation of the fourth Industrial Revolution and Industrie 4.0 (hereafter 4th IR), noting there is a linear and somewhat determinist trajectory flowing from the 1st IR to the 4th. In more sophisticated accounts this representation is seen as a simplification of what are complex processes. Yet diagrams similar to the one used in the previous chapter (see page 3) are mirrored in any number of policy document as well as in those prepared by consultants. Importantly, those who promulgate notions of the 4th IR are at one and the same time constructing and reconstituting a particular understanding of the future and importantly, the way it is conceived and understood. This is often an iterative and circular process with the same arguments being reproduced in a range of documents. Consequently, we encounter a number of recurring themes, robotisation, artificial intelligence (AI), digitalisation, smart factories, interoperability, interconnectivity, nanotechnology and so on. Such notions seep into popular culture and media representations and contribute towards popular and common sense understanding of the 4th IR and the socio-economic context in which it is set, one in which neo-liberalism is hegemonic (Hajkowicz et al. 2016; Bloem et al. 2014). These representations also draw upon science fiction, from Lang's classic film *Metropolis* to *Star Trek*. Channel 4, a British television channel, in 2017 had a Rise of the Robots season, which explored driverless cars and AI as well as sex bots, with BBC channel three presenting a programme entitled 'Can Robots Love Us?' (and see Dyer-Witheford et al. 2019: 24–28). While these latter illustrations are somewhat trite they anticipate fundamental societal changes which impact upon social relations, building on the transformative power of technology that may well exaggerate the potentiality of AI (Chollet 2017; but see Mason 2019). Such processes and cultural practices contribute towards the way in which we make sense of the 4th IR.

Social Imaginaries

Drawing on Taylor's (2002, 2004) terms, common sense understandings of the 4th IR contribute to the formation of a social imaginary as well as the manner in which this is reflected in social practices and their legitimisation. Taylor uses the notion of social imaginary to refer to the way in which 'people imagine their social existence' (2004: 23). To the extent that conceptualisations of the 4th IR flow from theorists to the elite and then to the many, these can be construed as a social imaginary. For Taylor social imaginaries are distinct from social theory as a result of the former's 'unlimited and indefinite nature' (25). Such imagining can inform the way in which we make sense of the 4th IR and the cluster of terms that surround it—robotisation, AI, big data and so on, as well as the contradictions and tension that surround these notions.

There are, however, other ways in which we could think about social imaginaries which draw on rather different but related understandings. For Taylor (24) social imaginaries often derive from the work of social theorists which infiltrate the imagining of the elite and then to wider society. Not dissimilar conceptualisations are found in Anderson's *Imagined Communities* (1983), Appadurai's notion of the imaginary as a social practice (1996), and Said's imaginative geographies (2003). All of these writers emphasise the play of social imaginaries in the field of practice and their emergent qualities, whilst playing down or ignoring the salience of science and technology (Miller 2015; Jasanoff and Kim 2015). Although Appadurai (1996: 34) discusses technoscapes which refers to 'the global configuration, also ever fluid, of technology and the fact that technology, both high and low, both mechanical and informational, now moves at high speeds across various kinds of previously impervious boundaries', there is a tendency to conceive of such flows as cultural and in this way to bypass socio-technical imaginaries. At the same time, he reminds us that 'imagination has become an organised field of social practices, a form of work (in the sense of both labor and culturally organized practice), and a form of negotiation between sites of agency (individuals) and globally defined fields of possibility' (Appadurai 1996: 31). The imaginaries surrounding the 4th IR would be a case in point, these can be construed as discourses which have material consequences. Watkins (2015: 599), referencing Said's (2003), spatial imaginaries suggests 'that these are socially held stories, ways of representing and talking about places and spaces', but in the context of this chapter the stories and representations refer to the economy and socio-economic relations.

Jasanoff (2015: 4) argues that socio-technical imaginaries are

> collectively held, institutionally stabilized, and publically performed visions of desirable futures, animated by shared understandings of forms of social life and social order attainable through, and supportive of advances in science and technology.

Much of the discussion surrounding the 4th IR is set within such an imaginary but is one that has material consequences for waged labour and vocational education and training (VET). Perhaps echoing modernity's 'consciousness' of society as an economy (Taylor 2002: 105). It is also set within attempts to secure consensual understanding of the future of society.

Pfeiffer (2017) explores the imaginary surrounding the 4th IR by drawing on Adam and Groves *Future Matters* (2007), as a future to be told, tamed, and traded. By these terms she has in mind the manner in which a discourse or imaginary is constructed around the 4th IR. In which the future is told, one marked by the velocity and rapidity of change that brings in its wake a series of disruptions that need to be tamed or managed (Oschinski and Wyonch 2017). At the same time this discourse constitutes a future to be traded and invested in. Importantly, these constructions are not restricted to the west but are also globally inflected. These imaginaries will be able to encompass not only the developed economies but also those which are emerging and underdeveloped, albeit that these imaginaries will be refracted accordingly. These imaginaries are in their different and sometimes contradictory ways attempting to be performative and to construct persuasive fictions set within a neoliberalism or post-neo-liberalism or minimally pro-capitalist stance.

Ideology

There is an affinity between the previous discussion and a whole raft of terms amongst which we could include hegemony, discourse as well as ideology. Althusser's structural Marxism offers a particular understanding of ideological relations and discourse that can be applied to an analysis of 4th IR and which complements the preceding discussion. Consequently, there are a number of ideas I would like to draw from his work.

Ideology is a representation of the imaginary relationship of individuals to their real conditions of existence. (1972: 264)
Ideology has a material existence. (1972: 267)
Ideology interpellates individuals as subjects. (1972: 270)

Ideology represents an imaginary of the relationship of the individual to the social formation and its socio-economic relations. In other words, a representation of individuals and their real conditions of existence. In addition the materiality of ideology informs social practices. Finally, ideology interpellates or positions the individual as a particular type of subject.

If we think back to the 4th IR and its socio-technical imaginaries, we can discern the way in which it represents the imaginary relationship of individuals to their real conditions of existence as well as the way in which this imaginary informs practice—its materiality. We can also discern the manner in which the 4th IR seeks to construct or interpellate the individual as a particular kind of subject.

Cohen (1972, 1997), whose discussion of youth subcultures was in part informed by Althusser, suggests, '*that the latent function of subculture is this to express and resolve, albeit 'magically', the contradictions which remain hidden or unresolved in the parent culture* (1997: 57). Perhaps we could conceive of the imaginaries that surround the 4th IR as mirroring this process, whereby they seek to overcome and magically resolve the tensions and contradictions surrounding the 4th IR and its broader socio-economic context. What then are these tensions and contradictions?

Tensions and Contradictions

- The polarisation of jobs, income and wealth (Citi GPS: Global Perspectives & Solutions 2016; Goos and Manning 2007; Goos et al. 2014; Autor 2015);
- A shift in favour of capital away from labour (Lordon 2014: 40);
- A tendency towards monopolisation and oligopolisation (Harvey 2014);
- The existence of technical unemployment, labour shortages and skill gaps (Citi GPS: Global Perspectives & Solutions 2016; Pew Research Center 2017);
- The contradictory existence of demographic changes benefitting either older or younger workers and or female workers (Goodhart and Pradhan 2017; MGI 2017; WEF 2016);
- Deskilling and/or upskilling (Alhammadi et al. 2014);
- A surplus population excluded from waged labour (Avent 2017);
- The growth of precariousness (Standing 2016);
- Ecological despoilation (Hajkowicz et al. 2016);
- Or indeed any combination of the above (Dolphin 2015).

The above could be construed as an arbitrary list of issues that we encounter in the current socio-economic conditions (Avis 2018). Yet they can all be related to aspects of the 4th IR. The polarisation of income and wealth can be linked to the significance of rentier capitalism as well as to the hollowing out of middling jobs in the labour market, both of which have been in part facilitated by digitalisation and the IoT. These changes themselves are reflected in the increasing power of capital in its relation to labour and the growth of precarious work. Yet at the same time these ideas are called to account by those writers who explore demographic dividends in relation to younger or older workers and by those who emphasise what may be described as a gender premium. That is to say, the argument that women are an underutilised resource that if fully mobilised could make a significant contribution to economic growth especially in the emerging and underdeveloped economies of the global south. The World Economic Forum (WEF 2016, unnumbered) states,

> Achieving gender equality isn't just a moral issue—it makes economic sense. Equality between men and women in all aspects of life, from access to health and education to political power and earning potential, is fundamental to whether and how societies thrive.

The move towards monopolisation and oligopolisation is reflected in the significance of platform capitalism and the power of Amazon, Google, Uber and Facebook which has been facilitated by digitalisation, the mobilisation of big data and the use of algorithms as well as by the logic of capitalism (Srnicek 2017). Mason (2019: 115–131) reminds us of the materiality of the processes involved in the storing of data and its dissemination as well as the environmental costs that are incurred. Yet within the socio-technical imaginary of the 4th IR there is the promise that whatever tensions or contradictions or difficulties that arise, these can be successfully managed for the benefit of all, or rather and more precisely, capital. Means (2018) draws on Morozov's (2013) notion of solutionism to describe such techno-determinist processes (and see McAfee). The tenor of such sensibilities can be seen in this lengthy passage from the WEF (2018a: 12, 13):

> There are a number of potential courses of action that governments, businesses and other actors could take to prepare today to enhance the likelihood of a world where people are able to engage in meaningful, fulfilling and safe work; where workers impacted by changes to the labour market are supported through transitions; where people create new opportunities for themselves and those around them; and where growth is human-centred. Some of these may be thought of as 'safe bets' or 'no regret' actions, in the sense that there is generally a strong consensus across society and different stakeholders concerning their desirability…
>
> Other measures could be conceived of as 'big bets'… but currently command significantly less consensus, particularly as they often entail different views about the meaning of work, the worth of technology and values around social equality. Therefore, which precise courses of action might be thought of in each way will depend on a range of circumstances and attitudes in each geography and society.

There are any number of points that could be made about this passage but I want to draw out some in particular that bear upon the future of work and its socio-economic context. Whilst the passage is drawn from a document that explored eight scenarios that consider the future of work it suggests that there is a consensus amongst key stakeholders, governments and businesses, concerning the development of meaningful and safe work and so on. It calls for human-centred growth. In such a context, workers are to be supported through the disruptions that derive from changes in the labour market, being placed within a context whereby they will be enabled to create opportunities for themselves—a call for the development of the neo-liberal enterprising subject. Second, it acknowledges a range of debates concerning work, technology, values and social equality. In addition, the same document points towards a number of 'non-exhaustive' actions that could 'proactively shape a better future' (WEF 2018a: 12–13), these are listed below.

- Workforce re-skilling
- Education systems reform
- Enhanced digital access

Re-skilling, educational reform and enhanced digital access are all concerned with the development of labour and the affordances that can be offered. In addition there is a call for another set of policy interventions to address the disruption that follows the velocity of change accompanying the 4th IR. These call for support of workers during moments of transition by providing income support, job protections, as well as encouraging 'smart' job creation and the development of mass entrepreneurship.

- Agile safety nets
- Job protection incentives
- Smart job creation incentives
- Support to mass entrepreneurship (WEF 2018a: 12–13)

There are also concerns about the governance of online platforms to ensure that fair work is provided. This sits alongside a concern with the mobility of labour whereby the movement particularly of skilled workers is facilitated across national borders. Finally, participation incentives will enable the development of flexible working arrangements that are to address both the needs of industry as well as those of workers who would benefit from part-time work.

- Governance of online platform work
- Mobility management
- Participation incentives (WEF 2018a: 12–13)

All these policy proposal work on a very particular terrain that takes for granted existing socio-economic relations, rooted in neo-liberalism and capitalist relations. In this sense, these arguments constitute a future not only 'to be told' but also 'to be tamed and traded'. In this instance, the radical affordances offered by the new conditions that surround the 4th IR are constrained by their location within existing socio-economic relations. Thus, the affordances offered by the 4th IR are tamed and consequently become domesticated by being tied to the status quo, having their potential radicalism undermined. It is important to acknowledge, lest we forget, that technology is never innocent (Avis 1981; Dyer-Witheford et al. 2019, 20). The affordances that derive from the technologies surrounding the 4th IR will have been shaped and enmeshed by capitalist relations and interests—therefore, it is necessary to go beyond these and transform them to benefit all. These technologies are a site of struggle and whilst they will have been shaped and bent to serve capitalism, there are nevertheless spaces that need to be searched out. These spaces can be occupied and appropriated to pursue progressive possibilities, albeit that this is a fraught and contradictory process.

Supranational Organisations

There is an affinity between the arguments proposed by the World Economic Forum and those of other supranational organisations. Such connections for Miller (2015: 277, 278) represent what he describes as a globalism in which technoscientific notions are put to work in institutions concerned with international governance. The debates surrounding the 4th IR would be a case in point as are the interventions made by the Commission of the European Union and its 'new' skills agenda (2016). The agenda suggests that skills lead to enhanced employability and prosperity for all stakeholders, workers, businesses and wider society. Notably, it mobilises a fairly truncated notion of skill as referring to 'what a person knows, understands and can do' (EU commission 2016: 2). Amongst these skills the digital are of prime importance. The digital transformation of the economy and of society is re-shaping not only labour processes but also the manner in which people interact and engage with one another. Consequently digital literacy is becoming a prerequisite for civic engagement. The EU commission (2016: 7) suggests that 'Europe needs digitally smart people who are not only able to use but also to innovate and lead in using these technologies. Without it Europe will not succeed in embracing this digital transformation'.

It is also important to acknowledge that in the new skills agenda there is a resonance with earlier social democratic concerns with social inclusion and cohesion as well as the European social market economy. Perhaps this suggests the EU commission has a specific relation to fractions of capital or to particular constellations of interest groups. A few years ago at a conference on inclusive capitalism, Mark Carney (2014), the then Governor of the Bank of England, and Christine Lagarde (2014), the then Managing Director of the International Monetary Fund, called for the development of a more inclusive capitalism. They both sought to mitigate the excesses spawned by finance capitalism whilst at the same time securing conditions favourable for the continuance of the free market. Carney argued in the wake of the 2008 financial crisis that bankers who had made fortunes during and after the crisis should in future face the full discipline of the market. This would then serve to undermine their previous privileges and contribute towards the rebuilding of social capital as well as trust in financial institutions in wider society. But in addition the disciplines of the market would increase the dynamism of the economy. In a similar vein Lagarde was concerned with the rebuilding of trust as well as arguing for a more equitable distribution of income and wealth. Notably, this concern was set in a free market capitalism but one that sought to avoid an in-built tendency to implode and to become 'its own grave digger'. Lagarde associates an inclusive capitalism with meritocracy, democracy and the salience of social capital and trust, but also competitiveness and the pursuit of growth.

Resonating with the previous discussion are the UN's (2015)[1] *Sustainable Development Goals* (SDGs), which have been subject to ongoing critique and discussion. Hickel (2015) for example, suggests that the concern with poverty reduction and the call for economic growth sidesteps the issues of inequality by avoiding the challenge of ensuring a fairer and more equitable distribution of resources, income and wealth. Similarly Swain (2018: 342) comments,

> The SDGs aim to eradicate poverty, establish socioeconomic inclusion and protect the environment. Disparaging analysis, suggests that there exists a potential inconsistency in the SDGs, particularly between the socio-economic development and the environmental sustainability goals… Critiques also raise questions on if and whether the SDGs can be measured and monitored.

A number of the SDGs can be addressed by technical and VET. However, I would like to draw the readers' attention to some of these that are strongly related to VET and social justice.

Goal 4. Ensure inclusive and equitable quality education and promote lifelong learning opportunities for all
Goal 5. Achieve gender equality and empower all women and girls
Goal 8. Promote sustained, inclusive and sustainable economic growth, full and productive employment and decent work for all
Goal 9. Build resilient infrastructure, promote inclusive and sustainable industrialization and foster innovation
Goal 10. Reduce inequality within and among countries

These goals emphasise gender equality, lifelong learning, economic growth, full employment, productive and decent work for all which is set alongside the reduction of inequality. These goals can be related to hegemonic debates surrounding the current economic conditions faced across the globe, mirroring Miller's notion of globalism. As such they resonate with those favoured by other supranational organisations. Unsurprisingly Guy Ryder, the Director General of the International Labour Organization reiterates these concerns in his forward to the ILO's (2017: v) *Inception Report: for the Global Commission on the Future of Work.*

> As the ILO approaches its centenary in 2019, the stakes are high—and intensifying—with respect to the importance of achieving social justice. Concern about economic growth and human progress is growing as inequality, insecurity and exclusion—which imperil lives and livelihoods and pose existential threats to social organization and democratic societies—become more widespread. Beyond 2019, the mandate to achieve the 2030 Agenda for Sustainable Development and its 17 Sustainable Development Goals (SDGs) adds urgency and opportunity to the ILO's tasks, and underscores the critical role of environmental sustainability in creating decent work and an inclusive global economy.

These themes are also taken up in UNESCO's *Strategy for Technical and Vocational Education and Training* (2016: 6) which has three priority areas:

(1) Fostering youth employment and entrepreneurship;
(2) Promoting equity and gender equality;
(3) Facilitating the transition to green economies and sustainable societies.

There is something of a circular argument that is taken up by a number of supranational organisation which are characterised by variations on a theme. Whilst, there might be different accenting of these they are nevertheless set within a particular zeitgeist. Azevêdo prior to his re-appointment as Director General of the World Trade Organisation stated, 'if members grant me the honour of serving a second term as Director-General that will remain my clear goal. Not to pursue trade for trade's sake, but as an essential means to support jobs, growth and development' (2017: online). He calls for the development of inclusive societies and economies where those who are 'left-out' are brought back into the mainstream with domestic policy particularly in the areas of education, training being of paramount importance for this agenda (Azevêdo 2018: online).

A New Social Contract

At this juncture, it is useful to draw together the themes that criss-cross the supranational organisation I have discussed. We encounter commitments towards: decent work, gender diversity and equality, lifelong learning, the development of inclusive and environmentally sustainable societies, as well as to human-centred growth. There are a number of ways in which we can make sense of these concerns. There is, however, a paradox in that these interests rest within what Bastani (2019) after Fisher (2010) describes as capitalist realism, that is to say, the suggestion that there is no alternative to capitalism and the free market. Consequently, all that can be done is to mitigate or soften the excesses of neo-liberalism which has reached its limits (Bastani 2019; Mason 2015, 2019). The World Bank (2019: 10) commenting on the excesses of neo-liberalism suggests 'the current social contract is broken in most emerging economies, and it is looking increasingly out of date in some advanced economies'. If this position was accepted it would sit alongside the call for a new social contract that would address the previous concerns and that would seek to restore the legitimacy of the economic system. Importantly this would be done in a manner that secured capitalist interests.

The *Global Commission on the future of Work* (2019: 11–13) calls for a new social contract that places a human-centred agenda at its heart. This contract is to be based on three pillars which call for increasing investment in people's capabilities, in the institutions of work and in decent and sustainable work. If addressed such practices would support economic growth, equity and sustainability. For the World Bank this new social contract seeks to enshrine, the social protection and inclusion of workers and aims to develop their human capital thereby enabling workers to participate in competitive labour markets. Many other writers have called for a new social contract. Raghuram Rajan (2019) a former governor of the reserve bank of India and chief economist and director of research at the IMF, discusses the three pillars that support wider society—the state, the market and community. When these pillars are out of kilter difficulties occur in the wider society. In the current conjuncture, the community has been weakened and is out of equilibrium with the state and market with all the difficulties that have been mentioned earlier. For Rajan it is necessary to restore this balance or social contract so that society is enabled to face a better future. This would necessitate not only holding back the excesses of the market but also of authoritarian populist and anti-democratic states. These arguments are echoed by McAfee (2019: 265) and his 'four horsemen of the optimist'—capitalism, 'tech' progress, responsive government and public awareness which when combined and balanced could secure the well-being of all members of society.

West (2018) points to a number of interventions that could serve to manage the transition to the new economy that is presaged by the 4th IR. He discusses a number of ways in which workers could be supported and offered social protection in these new conditions, amongst which he includes a universal basic income as well as income support for those transitioning to the 'digital economy'. In the UK, the Institute of Fiscal Studies' Deaton Review has launched a debate about inequalities in the UK in the twenty-first century. It comments on a number of concerns about those who are being left behind—the poor, young people, those with few qualifications, the regions as against the metropolitan centres. These processes are reflected not only in disparities in income and wealth but also in life chances and the standard of living and so on (Joyce and Xu 2019: 2). The debates I have examine in this section indicate a concern with the social and political implications of increasing levels of inequality in relation to life chances, income and wealth as well as being manifest in regional disparities and in inequitable standards of living.

The call for a new social contract can be seen as an attempt by these writers and some supranational and national organisations to put in place a new settlement that secures condition that enable continued capital accumulation whilst simultaneously softening the excesses of the free market. Stiglitz (2019: 31) whilst commenting on neo-liberal America, states, 'We have to construct a new social contract that enables everyone in our rich country to live a decent middle-class life'. We could also interpret this as an attempt to re-fashion a progressive neo-liberalism that draws on a politics of recognition in relation to diversity in terms of gender, race and sexuality, but which also works with a hierarchical conceptualisation of equality and meritocracy, aligning these with particular fractions of capital (Fraser 2019: 11–18). In this instance, the supranational organisations mentioned above operate as if they are a quasi-state that seeks to co-ordinate capitalist organisations and manage class conflict. In this way, they seek to secure capital's long-term future which cannot be done by individual capitalists alone with the 'state' having to discipline wayward fractions of capital (Fleming 2017; Miliband 1973; Poulantzas 1978). Such an analysis can only take us so far but does alert us to the ideological role played by these constituencies. It also points towards particular coalitions of interest that coalesce around specific fractions of capital, the membership of supranational bodies and their links with labour movements. This coalition represents a progressive current within capital but one that is inevitably compromised by the reality of that economic system.

It is as well to remember the brutalities of so much waged labour that occurs in current conditions. We only need to think about those employed by Sports Direct, Amazon as well as the sweatshops and factories across the developing world (Dyer-Witheford et al. 2019: 74). It is also important to consider the levels of youth unemployment across the developed and developing world. For many years this has been a pivotal concern of supranational organisations and national governments but the rates of unemployment remain stubbornly high. This exists in both developed and developing societies, alongside large pools of surplus labour which have been expelled from the workforce but which may no longer be visible in employment statistics produced by the state (Blanchflower 2019: 23–27). In addition, we need to recognise the contradictory stances adopted by supranational organisations—the IMF's involvement in structural adjustment programmes that sought to impose the discipline of the market on developing economies but which exacerbated poverty and inequality (Forster et al. 2019). The European Central Bank and EU imposed austerity on Greece which resulted in high levels of unemployment, the deepening of inequality and a declining standard of living for working people.

Sustainable Growth

The OECD's environmental outlook to 2050 (2012: online) illustrated a number of the problems that the world faces. The report addresses a number of themes—socio-economic development, energy and land use, climate change, and health and environment. It spells out the danger of inaction. Without steps to mitigate the impact of these developments the world will face insurmountable problems that will undermine long-term sustainability and lead to ecological despoliation—that is to say, the spectre of climate change running out of control. A number of reports produced by the World Economic Forum (2018b, 2019a, b) suggest the 4th IR can mitigate these dangers by offering technical solutions that serve the interests of all stakeholders from businesses to the communities and consumers they serve, whether this be in the global north or south—a win-win situation. Thus, the WEF (2018b: 5) points towards five cross-industry trends:

(1) Advanced remanufacturing: Innovation in physical and digital technologies is fast becoming an enabler for closing the loop with cost-effective returns processing, robotic disassembly and advanced material sorting. Connected devices feed information back to design and engineering to improve product durability and performance. A cluster of technologies—augmented workforce systems and cobotics, in combination with digital track-and-trace systems to manage reverse logistics—can drive substantial triple-bottom-line value [social, ecological and financial], boost a brand's reputation and mitigate growing supply chain risks.

The other four trends concern the use of:

(2) New green materials in relation to packaging electronics as well as alternatives to meat and so on;
(3) Advanced and precision agriculture;
(4) Increased factory efficiency as a result of automation, the IoT and shortening of supply chains;

(5) Traceability serving to confirm the sources of materials and so on, which would also enable reverse logistics and engineering.

The point is that these processes would serve to reduce energy consumption and be aligned with green technologies. The argument is that by drawing on digital, physical, biological and smart technologies the 4th IR can serve to increase competitiveness, contribute to the efficiency of regional economies as well as to environmental sustainability (WEF 2019b: 5). In addition, the 4th IR is also thought to lead to the development of new business models which build on the affordances that these offer, for example, reverse engineering, circular economic business models such as product life extension and so on. The WEF (2019a: 8–9) suggests that in this way the 4th IR can,

> For manufacturers, mak[e] sustainable manufacturing integral to their strategy can reduce costs and enhance profitability, drive business growth and be used to build trust with stakeholders. For governments, sustainable manufacturing can make a positive environmental and social impact that contributes towards the fulfilment of the UN's Sustainable Development Goals.

Notably, there is a happy coincidence between the interests of manufacturers, governments and wider society. This is set around an economics orientated towards competitiveness and growth. The pursuit of which is deemed to benefit all members of society whether in the global north or south. These arguments are focused on what WEF (2019a) describes as *Beacons of Technology and Innovation in Manufacturing*, and serve as lighthouses to illuminate the direction that could be followed by others. As with much of this discussion, workers are constructed as individualised human capital—a neo-liberal or capitalist construction (WEF 2019a: 14). However, the concern with addressing the UN's Sustainable Development Goals provides a resource that can be mobilised in the struggle for the development of socially just societies. This can serve as a mechanism against which capital can be held to account. However, as with the call for a new social contract, it is set on a very particular terrain that accepts capitalist social relation and the presumption of scarcity on which capitalism is predicated. McGrath and Powell (2016: 17) remind us that as against

> the rush of policymakers and international aid agencies to plan for green skills, green jobs and green economies. These are ultimately not transformative as they neither accord sufficient agency to individuals nor confront the fundamental tensions that lie at the heart of sustainable development. Nor is the vision of skills development contained within them genuinely transformative. Rather, the dominant approach to green skills sits comfortably alongside narrow notions of employability; deficit accounts of learners' failings to develop the skills necessary to be good consumer-citizens; a human capital set of assumptions that are blind to the role of institutions and political economy; and an account of work that is still fundamentally productivist.

McGrath and Powell are suggesting that the discourses that surround the concerns of supranational institutions with environmental and green issues are not pushed far enough. Consequently, their potential radicalism is truncated and becomes lodged within a productivist paradigm (Anderson 2009) that serves to occlude post-work analyses of VET. In addition, productivism fails to address those excluded from the formal economy as well as unwaged care workers. An expansive conceptualisation of VET would go beyond such limitations and is in part what Zancajo and Valiente (2019) have in mind when they call for the politicisation of TVET.

Conclusion

There are contradictions and tensions surrounding the rhetoric concerned with just and sustainable livelihoods and decent work which become apparent when set against the conditions in which many people labour. We could consider the progressive rhetoric that addresses these conditions as an ideology developed by the apologists of capital. Such a stance would overlook the complex politics surrounding such interventions. Rather than dismissing the policy rhetoric as a tool of capital to secure favourable condition it would be better seen as a resource that can be drawn upon in the struggle for a socially just world.

In this chapter, I have considered the social imaginaries and ideologies that surround the 4th IR. This allowed me to explore the constructions of the 4th IR and the manner in which it is lodged within capitalist relations. I examined the contradictions and tension that surround the 4th IR and which are recognised by those who seek to address the positive affordances that it offers, drawing on the arguments promulgated by various supranational organisations. Some of these envisage a new social contract that seeks to overcome the excesses of finance, rentier and free market capitalism. Fraser and Jaeggi when referring to the taming of capital used the term in a different way to Pfeiffer who used it to refer to the manner in which capital domesticated the affordance offered by the 4th IR to secure its interests. In contradistinction, Fraser and Jaeggi write (2018: 5), 'How "tamed" capitalism can be and still be "capitalism" is largely a semantic issue. At the same time, the excesses and threats posed by contemporary capitalism might give us pause [to ask] whether the idea of "taming" capitalism is still adequate'. In other words, together with Fraser and Jaeggi (2018) we should question whether a new settlement or social contract that can balance the interests of labour and capital in the manner of social democracy or egalitarian capitalism is a feasible political project (Kenworth 2004). This also poses several additional questions key amongst which is whether existing capitalism serves as a brake on the development of society. That is to say, towards one based on abundance as post-capitalists would suggest (Bastani 2019; Mason 2015). If we wish to work towards a socially just society, one in which we are enabled to express our species being, we need to adopt a politics orientated towards the transformation of society, one that goes beyond the gradualism of revolutionary reformism. In this struggle, we should push the affordances offered by the 4th IR to their limit and go beyond the exaggerated and overplayed rhetoric of the WEF which promises much that capitalism will be unable to deliver but which serves as a placatory politics.

Acknowledgement This chapter draws in part on my 'Socio-technical imaginary of the fourth industrial revolution and its implications for vocational education and training: a literature review', published in the Journal of Vocational Education & Training, 2018. 70:3, 337–363, https://doi.org/10.1080/13636820.2018.1498907.

Note

1. UN Sustainable Development Goals

Goal 1. End poverty in all its forms everywhere

Goal 2. End hunger, achieve food security and improved nutrition and promote sustainable

agriculture

Goal 3. Ensure healthy lives and promote well-being for all at all ages

Goal 4. Ensure inclusive and equitable quality education and promote lifelong learning opportunities for all

Goal 5. Achieve gender equality and empower all women and girls

Goal 6. Ensure availability and sustainable management of water and sanitation for all

Goal 7. Ensure access to affordable, reliable, sustainable and modern energy for all

Goal 8. Promote sustained, inclusive and sustainable economic growth, full and productive employment and decent work for all

Goal 9. Build resilient infrastructure, promote inclusive and sustainable industrialization and foster innovation

Goal 10. Reduce inequality within and among countries

Goal 11. Make cities and human settlements inclusive, safe, resilient and sustainable

Goal 12. Ensure sustainable consumption and production patterns

Goal 13. Take urgent action to combat climate change and its impacts

Goal 14. Conserve and sustainably use the oceans, seas and marine resources for sustainable development

Goal 15. Protect, restore and promote sustainable use of terrestrial ecosystems, sustainably manage forests, combat desertification, and halt and reverse land degradation and halt biodiversity loss

Goal 16. Promote peaceful and inclusive societies for sustainable development, provide access to justice for all and build effective, accountable and inclusive institutions at all levels

Goal 17. Strengthen the means of implementation and revitalize the Global Partnership for Sustainable Development. (UN 2015: 18)

References

Adam, B., and C. Groves. 2007. *Future Matters: Action, Knowledge, Ethics*. London: Brill.

Alhammadi, Y., E. Brynjolfsson, F. MacCrory, and G. Westerman. 2014. Racing with and against the Machine: Changes in Occupational Skill Composition in an Era of Rapid Technological Advance, In *Thirty Fifth International Conference on Information Systems*, Auckland 2014 1–17

Althusser, L. 1972. Ideology and Ideological State Apparatuses. In *Education: Structure Society*, ed. B. Cosin. Harmondsworth: Penguin.

Anderson, B. 1983. *Imagined Communities*. London: Verso.

Anderson, D. 2009. Productivism and Ecologism: Changing Dis/courses in TVET. In *Work, Learning and Sustainable Development*, ed. J. Fien, R. Maclean, and Man-Gon Park, 35–57. Dordrecht: Springer.

Appadurai, A. 1996. *Modernity at Large*. Minneapolis: University of Minnesota Press.

Autor, D. 2015. Why are There Still so Many Jobs? the History and Future of Workplace Automation. *The Journal of Economic Perspectives* 29 (3): 3–30. https://doi.org/10.1257/jep.29.3.3.

Avent, R. 2017. *The Wealth of Humans*. London: Penguin.

Avis, J. 1981. Social and Technical Relations: The Case of FE. *British Journal of Sociology of Education* 2 (2): 145–161.

———. 2018. Socio-technical Imaginary of the Fourth Industrial Revolution and its Implications for Vocational Education and Training: A Literature Review. *Journal of Vocational Education & Training* 70 (3): 337–363. https://doi.org/10.1080/13636820.2018.1498907.

Azevêdo, R. 2017. General Council: Appointment of the Director-General, speech, February 28. https://www.wto.org/english/news_e/spra_e/spra158_e.htm

———. 2018. Strength and Flexibility Are Essential for an Effective Trading System, Speech April 3. https://www.wto.org/english/news_e/spra_e/spra217_e.htm

Bastani, A. 2019. *Fully Automated Luxury Communism*. London: Verso.

Blanchflower, D. 2019. *Not Working: Where Have All the Good Jobs Gone?* Princeton: Princeton University Press.

Bloem, J., M. Van Doorn, S. Duivestein, D. Excoffier, R. Maas, and E. Van Ommeren. 2014. The Fourth Industrial Revolution Things to Tighten the Link between IT and OT, VINT Research Report 3. https://www.sogeti.com/explore/reports/vint-research-3-things—the-fourthindustrial-revolution/.

Carney, M. 2014. Inclusive Capitalism: Creating a Sense of the Systemic. Speech at the Conference on Inclusive Capitalism, May 27. Accessed May 29, 2014. www.bankofengland.co.uk/publications/Documents/speeches/2014/speech731.pdf.

Chollet, F. 2017. The Impossibility of Intelligence Explosion, Medium. https://medium.com/@francois.chollet/the-impossibility-of-intelligence-explosion-5be4a9eda6ec.

Citi GPS: Global Perspectives & Solutions. 2016. *Technology at Work V2.0: The Future Is Not What It Used to Be*. https://www.oxfordmartin.ox.ac.uk/publications/view/2092

Cohen, P. 1972. Sub-cultural Conflict and Working Class Community, Issue 2 of Working Papers in Cultural Studies, Centre for Contemporary Cultural Studies, University of Birmingham.

Cohen, P. 1997. *Rethinking the Youth Question*. London: Macmillan.

Dolphin, T., ed. 2015. *Technology, Globalisation and the Future of Work in Europe: Essays on Employment in A Digitised Economy*. London: IPPR. https://www.ippr.org/publications/technology-globalisation-and-the-future-of-work-in-europe

Dyer-Witheford, N., A.M. Kjøsen, and J. Steinhoff. 2019. *Inhuman Power*. London: Pluto.

European Commission. 2016. Communication from the Commission to the European Parliament, the Council, the European Economic and Social Committee and the Committee of the Regions: A New Skills Agenda For Europe. https://ec.europa.eu/transparency/regdoc/rep/1/2016/EN/1-2016-381-EN-F1-1.PDF.

Fisher, M. 2010. *Capitalist Realism, Is there No Alternative*. London: Zero.

Fleming, P. 2017. *The Death of Homo Economicus*. London: Pluto.

Forster, T., A. Kentikelenis, B. Reinsberg, B. Stubbs, and L. King. 2019. How Structural Adjustment Programs Affect Inequality: A Disaggregated Analysis of IMF Conditionality, 1980–2014. *Social Science Research* 80: 83–113.

Fraser, N. 2019. *The Old is Dying and the New Cannot Be Born*. London: Verso.

Fraser, N., and R. Jaeggi. 2018. *Capitalism: A Conversation in Critical Theory*. Cambridge: Polity.

Global Commission on the Future of Work. 2019. *Work for a Brighter Future*. Geneva: ILO.

Goodhart, C., and M. Pradhan. 2017. Demographics Will Reverse Three Multi-Decade Global Trends. BIS Working Papers, No 656, Bank of International Settlements.

Goos, M., and A. Manning. 2007. Lousy and Lovely Jobs: The Rising Polarization of Work in Britain. *The Review of Economics and Statistics* 89 (1): 118–133. https://doi.org/10.1162/rest.89.1.118.

Goos, M., A. Manning, and A. Salomons. 2014. Explaining Job Polarization. *The American Economic Review* 104 (8): 2509–2526. https://doi.org/10.1257/aer.104.8.2509.

Hajkowicz, S., A. Reeson, L. Rudd, A. Bratanova, L. Hodgers, C. Mason, and N. Boughen. 2016. *Tomorrow's Digitally Enabled Workforce: Megatrends and Scenarios for Jobs and Employment in Australia over the Coming Twenty Years.* https://publications.csiro.au/rpr/pub?pid=csiro:EP161054&

Harvey, D. 2014. *Seventeen Contradictions and the End of Capitalism.* London: Profile.

Hickel, J. 2015. Five Reasons to Think Twice about the Unsustainable Development Goals. https://blogs.lse.ac.uk/africaatlse/2015/09/23/five-reasons-to-think-twice-about-the-uns-sustainable-development-goals/

ILO. 2017. Inception Report for the Global Commission on the Future of Work. Geneva, International Labour Office. https://www.ilo.org/wcmsp5/groups/public/---dgreports/---cabinet/documents/publication/wcms_591502.pdf.

Jasanoff, S. 2015. Future Imperfect: Science, Technology and the Imaginations of Modernity. In *Dreamscapes of Modernity*, ed. S. Jasanoff and S.-H. Kim, 1–33. Chicago: Chicago, University Press.

Jasanoff, S., and S.-H. Kim, eds. 2015. *Dreamscapes of Modernity, Chicago.* Chicago: University Press.

Joyce, R., and X. Xu. 2019. *Inequalities in the Twenty-first century: Introducing the IFS Deaton Review*. London the Institute for Fiscal Studies.

Kenworth, L. 2004. *Egalitarian Capitalism.* New York: Russell Sage Foundation.

Lagarde, C. 2014. Economic Inclusion and Financial Integrity. Address to the Conference on Inclusive Capitalism, London, May 27. Retrieved from www.imf.org/external/np/speeches/2014/052714.htm on 29 May 2014.

Lordon, F. 2014. *Willing Slaves of Capital.* London: Verso.

Mason, P. 2015. *Postcapitalism.* London: Allen Lane.

———. 2019. *Clear Bright Future.* London: Allen lane.

McAfee, A. 2019. *More from Less.* London: Simon and Schuster.

McGrath, S., and L. Powell. 2016. Skills for Sustainable Development: Transforming Vocational Education and Training beyond 2015. *International Journal of Educational Development* 50 (2016): 12–19.
Means. 2018. *Learning to Save the Future*. London: Routledge.
MGI. 2017. *A Future that Works: Automation, Employment, and Productivity*. https://www.mckinsey.com/~/media/McKinsey/Global%20Themes/Digital%20Disruption/Harnessing%20automation%20for%20a%20future%20that%20works/MGI-A-future-that-works-Executive-summary.ashx
Miliband, R. 1973. *The State in Capitalist Societies*. London: Quartet.
Miller, C.A. 2015. Globalizing Security. Science and the Transformation of Contemporary Political Imagination. In *Dreamscapes of Modernity*, ed. S. Jasanoff and S.-H. Kim, 277–299. Chicago: Chicago, University Press.
Morozov, E. 2013. *To Save Everything Click Here*. London: Penguin.
OECD. 2012. Environmental Outlook to 2050: The Consequences of Inaction—Key Facts and Figures. https://www.oecd.org/env/indicators-modelling-outlooks/oecdenvironmentaloutlookto2050theconsequencesofinaction-keyfactsandfigures.htm
Oschinski, M., and R. Wyonch. 2017. Future Shock? *the Impact of Automation on Canada's Labour Market*, C D Howe Institute Commentary Number 472. https://www.cdhowe.org/public-policy-research/future-shock-impact-automation-canadas-labour-market
Pew Research Center. 2017. *The Future of Jobs and Jobs Training*. http://www.pewinternet.org/2017/05/03/the-future-of-jobs-and-jobs-training/
Pfeiffer, S. 2017. The Vision of 'Industrie 4.0' In the Making—a Case of Future Told, Tamed, and Traded. *Nanoethics* 11: 107–121. https://doi.org/10.1007/s11569-016-0280-3.
Poulantzas, N. 1978. *State, Power, Socialism*. London: New Left Books.
Rajan, R. 2019. *The Third Pillar*. London: William Collins.
Said, E. 2003. *Orientalism*. Harmondsworth: Penguin.
Srnicek, N. 2017. *Platform Capitalism*. Cambridge: Polity.
Standing, G. 2016. *The Corruption of Capitalism*. London: Biteback.
Stiglitz, J. 2019. *People, Power and Profits*. Princeton: Princeton University Press.
Swain, R. 2018. A Critical Analysis of the Sustainable Development Goals. In *Handbook of Sustainability Science and Research*, ed. W. Filho, M. Mifsud, and P. Pace, 341–356. Dordrecht: Springer.
Taylor, C. 2002. Modern Social Imaginaries. *Public Culture* 14 (1): 91–124.
———. 2004. *Modern Social Imaginaries*. Durham: Duke University Press.

UN. 2015. Transforming Our World: The 2030 Agenda for Sustainable Development A/RES/70/1. https://www.un.org/ga/search/view_doc.asp?symbol=A/RES/70/1&Lang=E

UNESCO. 2016. *Strategy for Technical and Vocational Education and Training (TVET), (2016–2021).* Paris: UNESCO. https://unesdoc.unesco.org/ark:/48223/pf0000245239.

Watkins, J. 2015. Spatial Imaginaries Research in Geography. *Geography Compass* 9 (9): 508–522.

WEF. 2016. *White Paper What are the 10 Biggest Global Challenges?* https://www.weforum.org/agenda/2016/01/what-are-the-10-biggest-global-challenges/

——— in Collaboration with the Boston Consultancy Group. 2018a. *White Paper Eight Futures of Work Scenarios and their Implications.* http://www3.weforum.org/docs/WEF_FOW_Eight_Futures.pdf

———. 2018b. White Paper, Driving the Sustainability of Production Systems with Fourth Industrial Revolution Innovation. http://www3.weforum.org/docs/WEF_39558_White_Paper_Driving_the_Sustainability_of_Production_Systems_4IR.pdf

———. 2019a. White Paper Fourth Industrial Revolution Beacons of Technology and Innovation in Manufacturing. http://www3.weforum.org/docs/WEF_4IR_Beacons_of_Technology_and_Innovation_in_Manufacturing_report_2019.pdf

———. 2019b. White Paper Shaping the Sustainability of Production Systems: Fourth Industrial Revolution Technologies for Competitiveness and Sustainable Growth. https://www.weforum.org/whitepapers/shaping-the-sustainability-of-production-systems-fourth-industrial-revolution-technologies-for-competitiveness-and-sustainable-growth

West, D. 2018. *The Future of Work.* Washington: Brookings Institution Press.

World Bank. 2019. *World Development Report 2019: The Future of Work.* https://www.worldbank.org/en/publication/wdr2019

Zancajo, A., and O. Valiente. 2019. TVET Policy Reforms in Chile 2006–2018: between Human Capital and the Right to Education. *Journal of Vocational Education & Training* 71 (4): 579–599. https://doi.org/10.1080/13636820.2018.1548500.

3

Robotisation, Artificial Intelligence, Employment and the Fourth Industrial Revolution

Abstract Many authors have argued robotisation will have a profound impact on employment leading to the replacement of routine manual jobs resulting in technological unemployment. This is set against those who draw on historical examples to suggest this assertion is mistaken. This chapter critiques these arguments which are distanced from the lived experience of workers in disadvantaged regions of Europe and the US. To comprehend these processes it is necessary to engage with the capitalist labour process, class struggle and rentier capitalism and how these articulate with class, race and gender, recognising their intersectionality. In addition, these processes need to be related to the socio-economic conditions experienced by low- and middle-income economies. The chapter concludes by considering the implications for vocational education and training.

Keywords Fourth industrial revolution • Robotisation • Artificial intelligence • Labour market • Rentier capitalism

J. Avis, *Vocational Education in the Fourth Industrial Revolution*,
https://doi.org/10.1007/978-3-030-52032-8_3

Introduction

This chapter explores a wide-ranging literature that addresses the relationship between robotisation, Artificial Intelligence (AI), the labour market and the labour process, considering how these processes impact on not only the global north but also the south. It relates these debates to the significance of rentier capitalism and the articulation of class, race and gender and their implication for vocational education and training (VET). By way of introduction the chapter considers various trajectories in which the fourth Industrial Revolution (4th IR) is set.

Trajectories

In the introductory chapter I described a particular trajectory followed by industrial revolutions, flowing from the first to the fourth. The first Industrial Revolution was related to the development of steam power, mechanical production and the railways, with the second being marked by the development of mass production and electrification. The third featured the emergence of personal computers, digitalisation and the Internet with the fourth intensifying these features. However, this is but one periodicy which could be set amongst a number of others. For example, Brynjolfsson and McAfee (2014) describe the shift from the first to second machine age, while other writers discuss a periodicy related to forms of capitalism where we encounter shifts from industrial capitalism towards late, cognitive and/or immaterial capitalism (Boutang 2011; Gorz 2010). Whereas Dyer-Witheford et al. (2019: 51) point towards a shift from industrial to cybernetic capitalism which incorporates a move from post-Fordism to AI capitalism as it 'actually exists now', anticipating the full development of AI capitalism as a general condition of production and means of cognition (Dyer-Witheford et al. 2019, Chapter 2). By this Dyer-Witheford et al have in mind the manner in which AI could serve as a general infrastructural condition of production in much the same way as electricity or transport systems. Thus AI would serve as a means of cognition available to all capitalists.

From a quasi-Marxist position Engeström's (2010: 191) model of historical modes of production develops the work of Victor and Boynton (1998: 233). In this case, there is a shift away from craft production, to mass production, lean production, mass customisation, co-configuration and ultimately to social production. Social production anticipates the socialisation of the means of production. Adler's paleo-Marxism would construe the move towards social production in Engeström's (2010) model as indicative of outcomes derived from the contradiction between the forces and relations of production that portend societal transformation (Adler 2006; Adler and Heckscher 2006; and see Avis 2016: 66–69). In addition, Adler has discussed the evolution of organisational paradigms towards the network, whereby there are links across internal and external boundaries (Bodrožić and Adler 2018: 8). This not only reflects Engeström's (2010) notion of co-configuration but also the horizontal and networked relations facilitated by the technologies associated with the 4th IR. The Internet of Things (IoT) and Internet of Services (IoS) offer affordances that can lead to new organisational forms. Benoliel, one of the experts canvassed by the Pew Research Center,[1] comments, 'Robots will enable the creation of new environments… [which] will disrupt the way we are used to work in centralized organizations and enable the emergence of a new type of horizontal organizations' (cited in Pew 2014: 58). For Bodrožić and Adler (2018: 8) the network is an organisational paradigm that serves to link and rationalise 'processes across internal and external boundaries'. In this case the enterprise has access not only to other organisations but also to other sources of expertise (Bodrožić and Adler 2018: 30), whereas the preceding corporativist paradigm reflected 'the multi-divisional mass-production corporation with strategic integration but operating autonomy in the [organisations] divisions' (2018: 7). Each organisational paradigm seeks to overcome the difficulties encountered by earlier forms but which in turn generate their own contradictions, with the direction of change towards the 'socialisation of production', which anticipates the development of post-capitalism.

The above discussion resonates with Bastian's argument (2019; and see Avis 2020a: 3–4), which operates with a much broader historical sweep than the previous trajectories. He describes three disruptions each of which resulted in a transformation of society, its social relations as well as the constructions of what it is to be human. For Bastian the 4th IR is located within the third disruption. The first disruption represented the shift from a nomadic existence towards a more settled life organised around agriculture. The second disruption refers to the first Industrial Revolution which ushered in what Brynjolfsson and McAfee (2011) refer to as the first machine age (2011) which again reshaped what it was to be human. The third disruption, or in Brynjolfsson and McAfee's (2014) terms the 'second machine age', anticipated the 4th IR, presaging the possibility of a society of abundance able to transcend the scarcity that was a feature of previous societies and upon which capitalism is predicated. This would herald the possibility of post-capitalist and post-productivist social formations (Anderson 2009; Mason 2015).

What these trajectories have in common is that they seek to develop an understanding of the changes impacting upon wider society as well as waged and organisational relations. They point towards what could be described as tendential processes as well as an acknowledgement of capital's endless search for new opportunities to accumulate capital and appropriate value—that is to say, to make money (Harvey 2014). These models offer heuristic devices but should not be seen as necessarily indicative of sudden ruptures or breaks, in as much as there is a need to acknowledge uneven development and the presence of various modes of production within any social formation, albeit that one may be more dominant than the others.

In some respects, the affordances offered by social production allied with digitalisation echo a re-imagining of craft production whereby boundaries between organisations and wider sources of expertise become more porous. In their different ways these various trajectories contribute towards the formation of socio-technical imaginaries that can be mobilised in counter-hegemonic struggles (see Streeck 2016; Wallerstein et al. 2013). Such a stance is more apparent in Adler's (2006) paleo-Marxism with its conceptualisation of the 'socialisation of production' and in Engeström (2010) notion of social production which offer progressive possibilities. Such notions are prefigured in Bastani's conceptualisation of the third disruption. However, such possibilities should not be conceived of in a determinist or evolutionary fashion but rather in relation to the balance of power between capital and labour, in other words the result of class struggle.

Not dissimilar arguments can be found in analyses that engage with post- and anti-capitalism that similarly herald the possibility of post-scarcity social formations (Anderson 2009; Mason 2015, 2019; Bastani 2019; and see Chap. 4). Importantly such critical work is at one remove from discussions forwarded by the World Economic Forum, business consultancies and various supranational organisations that consider the affordance offered by the 4th IR but at the same time note the dislocations that might arise. There are at least two points to be made. Firstly, conceptualisations of the 4th IR are contested, albeit that some seek to gain a hegemonic position. In addition there is a tendency for discussions to veer towards a form of technical determinism whereby there is an inevitable trajectory that societies need to follow, that is, if they are concerned with the well-being of all. This tendency can be seen in accounts that reify the 4th IR, treating it as a social fact rather than as an ideological or imaginary construction. Yet at the same time we must acknowledge that these constructions of the 4th IR have material consequences, being a resource that can be drawn on in class struggle.

Robotisation

Any number of papers and books have argued that the growth of robotisation will have a profound impact on employment and will lead to the replacement of routine manual jobs by robots (Brynjolfsson and McAfee 2011, 2014; Ford 2016; Peters 2017). In a much-quoted paper, Frey and Osborne (2013: 21, 22), citing a technical report produced by McKinsey Global Institute (MGI), write,

> Estimates by MGI (2013) suggests that sophisticated algorithms could substitute for approximately 140 million full-time knowledge workers worldwide … The trend is clear: computers increasingly challenge human labour in a wide range of cognitive tasks.

Such trends, or so it is claimed, will lead to technological unemployment and be exacerbated by developments in AI, the use of big data as well as the IoT. In the latter case interconnected sensors within a particular factory, but also across supply chains, will automatically determine levels of production without human intervention. This is the spectre of 'smart' factories, fully automated systems which have no need for waged labour, that transfer data through the Internet. There are at least two points to be raised. Firstly, many authors fears about the impact of technological change on employment are by no means new, having been expressed in the work of Marx (1976 [1867]), Braverman (1974) and many others (see Anslow 2016). Importantly, we need to acknowledge that technology is never neutral but is entwined with social relations (Mishel and Shierholz 2017). We can also think back to the Luddites (Bregman 2016: 75, 87–90; Sale 1996), to Keynes' (2009) intervention in the 1930s as well as to the worries over automation and employment that were raised in the 1950/1960s (Autor 2015: 3; Mokyr et al. 2015). Such worries have been shown to be misfounded (Teixeira 2017). Secondly, Baxter et al. (2012) and Pfeiffer (2014, 2016) have illustrated empirically the need for skilled labour to monitor and troubleshoot not only in advanced manufacturing but also in other automated processes (see Saniter and Howe 2017), such as in piloting a plane or in monitoring a chemical or nuclear plant. Skilled workers who formerly worked on machines that are now linked together through sensors will have a feel for the job. In Sennett's (2009) terms, such workers will possess tacit knowledge, having a material consciousness of the manufacturing process. Intuitively, from the sound of the machines, vibrations and so on the worker will gain a sense of what is right and wrong and therefore the adjustments that need to be made. Much the same point could be made about a skilled car mechanic who may, rather than relying on computerised sensors, prefer to listen to the sound of the engine. As automation advances and increasingly covers a range of interconnections and as workers become progressively distanced from working on machines to monitoring them, the level of skill rather than decreasing will increase in line with the complexity of the production process. A single worker may well be responsible for a number of interconnected machines, whereas previously there would have been a one to one relationship. In order to troubleshoot the worker will require a sophisticated understanding of production. Baxter et al. (2012: 66) remind us of one of the ironies of automation whereby if a system frequently crashes operators will become skilled in responding to these routinely. However, systems that rarely trigger alarms will require higher levels of training to ensure operators respond appropriately. At the same time, there is the danger of intense boredom leading to apathy of the kind Paulsen (2014) describes as a feature of 'empty labour' and which might be exacerbated by advances in AI. This applies particularly to those workers who have a strong sense of work obligation and find themselves in effect with little to do. It is also important not to exaggerate the impact of AI. Chollet (2017) reminds us that the enactment of intelligence is socially situated and consequently is more sophisticated than computational power alone.

The Labour Market

It is necessary to acknowledge that whilst the 4th IR is an ideological and imaginary construct, the technologies surrounding it—digitalisation, robotisation, AI, IoT, IoS and so on—will impact upon the labour market, the labour process and of course technical and vocational education and training (TVET). However, these material consequences will be thoroughly social, being shaped by the balance of power between labour and capital, the dominant forms of capital, geographical location and the specific competitive strategies of the particular organisation.

There is a raft of work that addresses the relationship between levels of employment and automisation and digitialisation with much of this literature suggesting the growth of technological unemployment (Avent 2017; Livingston 2016; Rifkin 1995, 2014; Ross 2016: 35–43). A far more cautious stance is adopted by labour economists who question the assumption of technological unemployment. Goos and Manning (2007) suggest that since 1975 the UK labour market has witnessed job polarisation with increased employment found amongst the highest and lowest waged occupations. A similar argument in relation to job polarisation in the US has been developed by Autor et al. (2006). Goos and Manning suggest that this is also a feature of the European labour market based on evidence drawn from 16 West European countries in the period 1993–2010 (Goos et al. 2014). However, in an analysis of job polarisation in Europe Cirillo (2018) offers a more nuanced approach. She sees job polarisation as being influence by sector as well as by its societal or spatial location, with this process being most marked in the UK. Job polarisation is a consequence of the hollowing out of the employment structure and the significance of non-routine service-level jobs. Goos, Manning and Salomons argue, drawing upon Autor's work, that analyses based upon 'skill biased technical change' can only take us so far in explaining these trends and that such arguments need to be allied with what they describe as Autor et al.'s (2003), 'routinisation' thesis. Writing in 2015, Autor (12) identifies two types of task that are difficult to computerise: first, those that are characterised by problem-solving, intuition, creativity and persuasion and, second, non-routine tasks involving situational adaptability, personal interaction which are features of occupations, such as

> food preparation and serving jobs, cleaning and janitorial work, grounds cleaning and maintenance, in-person health assistance by home health aides, and numerous jobs in security and protective services … While these activities are not highly skilled by the standards of the US labor market, they present daunting challenges for automation. (12)

Processes of job polarisation are also related to the specificity of the particular labour market in which they are set. There are two points to make. For example, an increased demand for workers at the top of the income structure can be allied with an increased demand for particular types of service jobs, janitors, cleaners and so on, which while remaining low waged may nevertheless gain increased income compared with other similarly placed low-waged workers. We could compare the features of the labour market found in the US rust belt with those of New York. In the latter there will be a demand for service workers which will be depressed in the former. Second, many middling level jobs will have routine features that render them amenable to computerisation which results in the hollowing out of occupational structures. We need to be aware of the subtleties surrounding the debates that consider processes of job polarisation and substitution in order to assess which low-waged or middle-level occupation is subject to these processes. The significance of low-waged service workers has already been considered but much the same argument could be directed at middle-level occupations. Holzer (2015) suggests that many economists emphasise the role of digital technologies in hollowing out middle-skill jobs, as against this he argues that this applies to some but not all such jobs (and see Susskind and Susskind 2016). There are 'middle-skill' jobs that require post-secondary education, for example, health care, mechanical maintenance and repair, that are not amenable to automation. In addition, in the case of some of these jobs employers find it difficult to recruit labour. This may be partly a consequence of the relatively low wages offered and the reluctance of employers to increase these as they fear this would reduce their competitiveness. Autor (2015: 26) speculates,

> Employment polarization will not continue indefinitely (as argued in Autor 2013). While some of the tasks in many current middle-skill jobs are susceptible to automation, many middle-skill jobs will continue to demand a mixture of tasks from across the skill spectrum. For example, medical support occupations—radiology technicians, phlebotomists, nurse technicians, and others—are a significant and rapidly growing category of relatively well-remunerated, middle-skill employment.

The hollowing out of middle-level jobs allied with the impact of digitalisation and robotisation upon routine level jobs has been associated with the growth in technical unemployment. However, many labour economists doubt the veracity of this argument. To secure their claim they draw upon the history of employment from the Luddites to the present day. Haldane (2015: 7) refers to the 'lump of labour' fallacy, the notion that technological change leads to increased levels of unemployment (for a not dissimilar argument, see Atkinson and Wu 2017). Haldane (2015) argues that one of the consequences of technology displacing labour is that goods and services become cheaper which then boosts demand for these and the new industries that supply them, which in turn leads to increased employment. He acknowledges that the balance between the loss of jobs as a result of technological substitution and compensatory processes leading to increased employment as result of the production of new goods and services is an empirical question. He nevertheless argues that since 1750 the direction of travel has been counter to technological unemployment. Rather he argues (2015: 9),

> Viewed over the sweep of history, then, there is essentially no evidence to suggest technology has damaged jobs and plenty to suggest it has boosted wages. Technology has enriched labour, not immiserated it. Mill was right; Ricardo was wrong. Labour is not dead wood to be carved up between tasks. It is a tree whose trunk and branches have lengthened and thickened with time. The "lump of labour" fallacy is just that.

This is a stance shared by the Institute of Directors. While technological innovation can disrupt labour markets it is suggested that claims of technology-induced mass unemployment have been misplaced with more jobs being created than lost (Nevin 2016: 9). Mishel and Bivens (2017) endorse this argument drawing on Acemoglu and Restrepo's (2017) analysis of the US economy. Similarly, Dittrich in a footnote (2016: 6 footnote 18) has challenged the belief that many well paid middle-level jobs will be lost as a result automation, using as evidence a survey of German companies which found that these employers planned to hire more workers rather than reduce employment (Seeverband der Bayerischen Metall- und Elektro-Arbeitsgeber 2016).

In many respects the previous arguments are convincing. The convictions of technological pessimists seem to be challenged by the optimists, that is, if we consider paid employment as a good in itself. Perhaps it would be salutary to recall Fleming's (2017: 5) excoriation of much theoretical and applied economics that purveys a narrative 'that defies the daily experience of the 99%-ers as the backwash of debt, stress and insecurity threatens to drag us under'. We could set Fleming's comment against the frequently repeated rather more optimistic accounts that celebrate the growth in employment in the UK (Coyle 2017; DWP 2019; McRae 2019).

There are, however, a number of issues that need to be addressed. While historically technologically determined mass unemployment has not been a permanent feature of social formations in the west, it does not necessarily follow that this will be the case in the future. Autor (2015: 4) writes,

> Clearly, the past two centuries of automation and technological progress have not made human labor obsolete: the employment-to-population ratio rose during the twentieth century even as women moved from home to market; and although the unemployment rate fluctuates cyclically, there is no apparent long-run increase. But those concerned about automation and employment are quick to point out that past interactions between automation and employment cannot settle arguments about how these elements might interact in the future: in particular, the emergence of greatly improved computing power, artificial intelligence, and robotics raises the possibility of replacing labor on a scale not previously observed. There is no fundamental economic law that guarantees every adult will be able to earn a living solely on the basis of sound mind and good character.

We cannot view the past unproblematically to predict the future. In addition, we need to consider the nature of employment, its relationship to capitalism and the wider global economy. It is important to consider the articulation of these processes to the geographical formation of labour markets in both the global north and south. For example, Rodrik (2016: 2–3) has discussed 'premature deindustrialisation' in low- and medium-income economies, many of which will be located in the global south. He is using the term 'premature deindustrialisation' to suggest that these societies are undergoing deindustrialisation at a much earlier point than the historical norm found within the advanced economies of the west. Consequently, these late industrialisers are unable to build large manufacturing sectors and deindustrialise at lower levels of income than those societies that industrialised at an earlier date. In addition, because of labour-saving technologies in the advanced economies, this opens up the possibility of reshoring production and the re-bundling of processes that were formerly separate. This is facilitated in part by the IoT with elements of production being automated through the use of the Internet and sensors (Livesey 2017; Whittaker 2015: 37). Such processes are both gendered and ethnicised. We need only think of the involvement of women drawn from the global south in care and service work in the north. In addition such processes are also found in the global north whereby ethnic, classed and gendered processes articulate with the labour market.

One of the successes of neo-liberalism which has impacted upon labour markets has been to break down national borders whilst simultaneously emphasising the importance of regionalism and localism. This can be seen in England in the division between north and south, with the former having higher rates of disadvantage in terms of unemployment and restricted labour markets. Notably, such regional disparities are also found in continental Europe, East and West Germany being a case in point (Roberts 2016; Solow 2008). Social geographers Martin and Morrison (2003, and see Avis 2018) have drawn our attention to the spatial and constructed nature of labour markets as well as their porosity. Alongside a local labour market of the type that Shildrick et al. (2012) suggest exists in the north of England that features low-waged intermittent work, or indeed no work at all, there may be other workers lodged within a global labour market of high-skilled/waged work. Such global and local labour markets may in some senses overlap but will also be ongoingly constructed and subject to change (Martin and Morrison 2003). This means that within a social formation, areas of full employment and putative skills gaps/mismatch sit alongside regions/localities characterised by multiple disadvantage and the lack of decent jobs (Keep and James 2012; Orr 2016). These processes are not only classed but also gendered and ethnicised, as well as being a site of struggle. In the global south not dissimilar processes can be mapped on to what Brown et al. (2011: 64) refer to as 'oasis operations' whereby hi-tech facilities can be set in 'low-spec' neighbourhoods in which they have little or no impact on the local labour market.

It is also important to acknowledge intersectionality—that is the manner in which class, ethnicity, gender and so on articulate to one another as well as the way in which these are mediated in specific social formations. For example, gendered, classed and ethnicised relations will be accented differently in Sweden, England and South Africa and it is important that analyses recognise this. At the same time we must acknowledge the way in which the technologies associated with the 4th IR are mobilised as part of a firm's competitive strategy. If it is cheaper to employ waged labour than to replace it with machines, the former may be chosen even though it may be less efficient and use outdated technology.

Rentier Capitalism

Workerist analyses have suggested that the shift from Fordism to post-Fordism was capital's response to class struggle and the refusal to labour. Lotringer (2004: 11) writes, 'It was Italian workers' stubborn resistance to the Fordist rationalization of work… that forced capital to make a leap into the post-Fordist era of immaterial work' (2004: 11). Regardless of the veracity of this claim in relation to the immateriality of the knowledge economy and in particular the failure to recognise its underpinning materiality, we need only think about the material infrastructure that supports the internet, super-cooled servers, cabling and so on (Mould 2018: 135). Workerist arguments also address the way in which surplus value and profits are accumulated. There are two key points to be made; the first concerns the manner in which surpluses arise when 'all of life is put to work'. We can think about our engagement with the internet and the forms of income secured by online platforms as a result of selling data and gaining advertising revenue. In this instance, our leisure activities create value for online platforms, such as Google. Second, in the current conjuncture rent seeking has both become an important means of creating profits but also one that is increasingly facing criticism. In recent years in many of the social formations in the global north, financialisation and rentier capitalism has assumed greater significance, this being particularly the case in the Anglophone societies. There are several consequences that flow from this, key amongst which is the shift in power from labour to capital. In rentier capitalism the aim is to generate profit whilst avoiding an engagement with production—the pursuit of primitive accumulation. In other words, money is used to create money, that is to say 'M-M' without any investment in the means of production. The consequence is that this will ultimately lead to a declining rate and crisis of profitability.

Wolf (2019: online), an apologist for capitalism reflecting on the failures of rentier capitalism, writes:

> We need a dynamic capitalist economy that gives everybody a justified belief that they can share in the benefits. What we increasingly seem to have instead is an unstable rentier capitalism, weakened competition, feeble productivity growth, high inequality and, not coincidentally, an increasingly degraded democracy. Fixing this is a challenge for us all, but especially for those who run the world's most important businesses. The way our economic and political systems work must change, or they will perish.

As with the discussion of the limits of neo-liberalism in the preceding chapter, there is increasingly a recognition that the current form of capitalism is in need of revision so as to ensure its continuance. There are a number of issues that flow from this that have a bearing on vocational education and training amongst which we could include questions of flexibility, adaptability and perhaps most importantly worklessness as well as and relatedly exclusion from waged labour. Holloway (2019: 230) suggests the call for more, better or less alienating waged labour only serves to lock us into the capitalist system. Cleaver (2017: 13) echoes this noting the 'centrality of the struggle against work is the most fundamental threat to the [capitalist] machine'. There are several ways in which this argument could develop. It could be argued that vocational education and training is so enmeshed with capitalism that it is ultimately a conservative force. Echoing such a stance Allman et al. (2003: 149–150) argue,

> Education [and we might add VET] plays a key role in the perpetuation of the capital relation; this is the skeleton in capitalist education's dank basement. It is just one of the many reasons why, in contemporary capitalist society, education assumes a grotesque and perverted form. It links the chains that bind our souls to capital.

However, an alternative and more nuanced account could be developed. Such an account whilst acknowledging the manner in which VET is enmeshed with capitalist relations simultaneously considers the radical affordances provided by it, recognising the surrounding contradictions and tensions. How then should we think about VET, not only in conditions of uncertainty but also in those of worklessness, so as to anticipate the transformation of society? This would call for a rights-based, post-productivist conceptualisation of VET predicated on the development of a socially just social formation (Anderson 2009; McGrath and Powell 2016)

Vocational Education and Training

Whilst the 4th IR is a rhetorical construction it nevertheless points towards socio-economic and technological changes that have impacted upon the labour market and by default VET. It is worthwhile reiterating that technology is never innocent as it is always embedded in social and economic relations. Much the same can be said about education and VET both of which are placed in a specific context in both the global north and south. This is a context in which whilst capitalist interests are hegemonic but one in which they are also open to contestation. Consequently any discussion of VET is Janus-like, facing in contradictory directions simultaneously. The forms of knowledge collectively developed or indeed the skills that are honed through participation in communities of practice, not only open up but also close down wider and particular understandings of the nature of work in a capitalist society as well as its possible transcendence. An individual and collectives sense of occupational or professional identity may provide a resource that can be used to critique the labour process and the practices of employers, yet may also emphasise the responsibilities owed to the latter. The different socio-economic, national, regional and global contexts in which VET takes place as well as the level at which it is set may lead to narrower or broader conceptualisations of the resulting education and training. We can think about the way in which conceptualisations of competence are mobilised in the Anglophone societies of the US, Australia, the UK and so on, setting this against the manner in which it is understood in Germany and much of continental Europe. Winch writes that the German notion of Kompetenzen embodies a broader understanding of competence than that found in the Anglophone societies.

> [Kompetenzen] is more than a bundle of skills, but is unified through a conception of agency which involves planning, control, co-ordination, self-monitoring and evaluation, as well as the performance of a variety of tasks requiring specific skills. It also includes the ability to appreciate the broader economic and civic implications of occupational action. (Winch 2012: 179)

Notably, this includes 'the ability to appreciate the broader economic and civic implications of occupational action' which goes beyond Anglophone, restrictive and narrow definitions of competence and serves as a critical resource that can call not only practitioners to account but also the organisations and social formations in which they labour. Yet at the same time, the notion of Kompetenzen rests within a specific socio-economic formation that accepts capitalist relations (see Avis 2010; Wheelahan 2019), albeit one in which the excesses of neo-liberalism are softened. Inevitably, and as with any form of educational practice, the question of pedagogy arises. In some ways it would be risible to suggest a distinctive pedagogy that is appropriate for the 4th IR. Hodkinson and James (2003: 400–401) suggest:

> What counts as good teaching and learning can be very different from site to site, and the complex interactions between numerous factors, almost inevitably means that specific characteristics of any site should be recognised and taken into account. It is this fact that leads to our first two major interim findings:
>
> 1. *There are considerable benefits in understanding learning as a cultural phenomenon.*
> 2. *What works, or is deemed good practice in one learning site may not work or be good practice in another.*
>
> Both these findings are strongly at odds with the current hegemonic rhetoric about FE [VET], based as it is on the idea that individual needs are sufficiently known to tailor provision most effectively, within a framework of standardised approaches to teaching, measuring and funding success.

Hodkinson and James remind us of the specificity of the teaching context not only in terms of its cultural location, which is in part constituted by pedagogic practice, but also indicates the fluidity of what constitutes good practice. A lecture may be seen as an example of 'banking' educational practice as it positions learners as passive recipients of knowledge. The lecture here is constituted as a 'readerly' text that offers an authoritative account. Yet some lectures may embody aspects of co-production between participants and lecturer, becoming in effect 'writerly' texts which open up dialogue and which do not close down a range of readings (Ball et al. 2011:15). Such a practice would serve to problematise authoritative texts, opening them up to critique and would engage with learners in developing evaluative tools. There is a connection here with curriculum debates that engage with disciplinary knowledge. Bernstein (1977: 97–98) reminds us,

> The ultimate mystery of the subject [discipline] is revealed very late in the educational life. By the ultimate mystery of the subject, I mean its potential for creating new realities. It is also the case, and this is important, that the ultimate mystery of the subject is not coherence, but incoherence: not order but disorder, not the known but the unknown. As this mystery... is revealed very late in the educational life... only the select few experience in their bones the notion that knowledge is permeable, that its orderings are provisional, that the dialectic of knowledge is closure and openness.

There are several points to be made. Whilst disciplinary knowledge is provisional and contestable, it also carries with it epistemic gains. This is because it provides access to conceptualisations that help us to understand both the natural and social world, that are not readily available to vernacular knowledges. It is for this reason that the disciplines provide us with powerful knowledge (Wheelahan 2019; Young 2008). However, importantly we need to acknowledge their social construction and provisionality, consequently in order to do this we need to have access to the appropriate evaluative tools. That is to say, we need to be appraised of the ways in which such knowledge claims are justified and legitimated, for without this we will be badly placed to evaluate and critique them. This is a matter of distributive justice (Wheelahan 2019). Narrow conceptualisation of VET that draws on disciplinary knowledge in a piecemeal fashion often fails to provide learners with the means to critique this knowledge. By default such conceptualisations of VET lead to truncated and restricted models of professionalism and what it is to be a member of a specific occupation or profession.

The preceding discussion raises a number of salient points. Such discussions could be characterised as distanced from occupational and professional practice in that they are over intellectualised and reflect a drift towards general education, moving away from a direct engagement with the world of work (Guile and Unwin 2019a: 5). Thus, whilst the above may be applicable to academic education it underplays the necessary relationship of VET to practice. However and importantly, some sort of notion of theory will nevertheless underpin occupational and professional practices. This becomes increasingly apparent if debate is shifted away from the notion of skill, which can so easily become enmeshed with narrow and restrictive conceptualisations of competence, towards one that is focused on expertise (Guile and Unwin 2019b). In some ways the notion of expertise sidesteps analyses that seek to list the types of skill that are required in the digital age or the knowledge economy, or indeed the current conjuncture. This is because the notion of expertise not only encompasses such skills but moves beyond them in that it implies a dialogic relationship between theory and practice. Expertise necessarily involves some sort of theoretical understanding of practice allied to its continuous development. It thereby provides a space for disciplinary knowledge as this both informs and is informed by practice—the two being entwined.

The notion of skill is often individualised and associated with the acquisition of human capital and qualifications, with the latter seen as a proxy for the former. However, expertise goes beyond the individual and arises through collective engagement with a community of practice. Consequently, and for Guile and Unwin (2019b: 29), one of the features of expertise is that it is '*Collective*: residing in teams, relational, co-produced, distributed'. It is also '*Practice-based:* involving individual and collective honing of skills. It does not necessarily mean there is an achievable point…' (2019b: 30). Expertise is also situational, contingent and continuously developed, involving the exercise of judgement and 'connoisseurship'. It will have been developed over the years through participation in a community of practice, whereby the practitioner has a deep and almost intuitive understanding of the field in which they are located and could be described as a form of 'connoisseurship'. Yet expertise will be subject to the limits and possibilities surrounding the context in which its enactment occurs, one that may be more or less expansive or restrictive—in the latter instance the moulding of VET to serve particular and performative interests would be a case in point.

The current conjuncture is not all of a piece, with different sectors of the economy being characterised by varying modes of production. The notion of uneven development captures some of these features but at the same time suggests a linear trajectory rather than the co-existence of different modes of production within any specific social formation, whether it be in the global north or south. It is important to acknowledge this co-existence as this may be overlooked or played down in discussions of the 4th IR and digitalisation. This is because the 4th IR can be seen as anticipating the transformation of economic and social relations. Such a stance carries with it the suggestion that other modes of production are of the past and ripe for transformation. Such a characterisation is clearly wrong. We need only reflect on various forms of care work which address individual well-being that are not readily amenable for digitalisation. We could also consider the current interest in the reinvigoration of craftwork in the global north, with the development of niche small scale artisan production. This also has a bearing on how we reflect on the notions of occupation and profession in the current conjuncture.

Much is made of the impact of technological development and digitalisation on professional and occupational groups. It is important to recognise that this is an uneven process. The use of AI, for example, has changed accountancy as well as the legal profession placing a greater emphasis on the development of 'soft' skills, the ability to communicate, persuade, problem solve, interpret data and so on (Ireland 2018; Susskind and Susskind 2016). This has resulted in analyses that consider occupations or professions as embodying a number of tasks some of which will be amenable to digitalisation and the use of AI, whereas other will not. This leads to the disassembling and reconfiguration or re-bundling of the tasks performed by these vocational groups. Such processes may at one and the same time undermine or enhance the expertise of the occupational or professional group, resulting in de-, re- or up-skilling. However, the way in which this is played out in practice is shaped by the balance of power between stakeholders some of whom are better able to shape the way in which technology is mobilised. This raises the question of power and the manner in which these processes articulate to capitalist relations, organisational strategy as well as the outcome of the struggles that may arise between particular constituents—capital, labour, the state and so on.

It is relatively easy to become caught up in the rhetoric that surrounds the 4th IR whether this is marked by pessimism or optimism, noting that these attributions are shaped by our positionality and are also the outcome of struggle (Avis 2020b). After all one person's pessimism may well be another's optimism. If we consider the 4th IR as indicative of the technical possibilities that surround the current conjunctural moment, we can avoid the exaggerated references to the distinctiveness of the 4th IR. Such a stance allows us to consider the affordance that the current conjuncture opens-up for the development of socially just social formations—a challenging project and one that faces innumerable obstacles. Perhaps what is new is the prospect of a society of abundance or at least one in which we can produce 'more from less' (McAfee 2019) (Rifkin 1995, 2014). Such developments would serve to recalibrate waged labour, production and the meaning of vocation as well as VET. We need to develop a rather more expansive notion of VET that can incorporate activities that lie outside the wage nexus, practices that are nevertheless productive but which contribute to collective and individual well-being. In other words, through 'really useful labour', forms of expertise could be developed which conjoin theory with practice and benefit wider society (Avis 2020a). Needless to say, such practices raise pedagogic and political questions and need to be aligned with a politics that seeks to develop a better world. It is doubtful whether capitalism would be able to deliver such a prospect though there are apologists that would argue that it would be able to do so. I have considered some of these arguments when I explored the call for a new social contract earlier in the book. McAfee (2019: 3–4) suggests that capitalism aligned with technological progress could lead to improvements in the human condition and the state of the natural world. That is, if capitalism and technological progress are aligned with the other two 'horsemen of the optimist' (4), public awareness and a responsive government. There is a space here in which some of the excesses of 'neo-liberalism' and rentier capitalism could be softened. However, I am reminded of the Fraser and Jaeggi (2018: 5) comment on 'How "tamed" capitalism can be and still be "capitalism"... at the same time, the excesses and threats posed by contemporary capitalism might give us pause over whether the idea of "taming" capitalism is still adequate'.

Conclusion

This chapter has explored a wide-ranging literature that has addressed the relationship between robotisation, AI, the labour market and the labour process and has referred to the manner in which these processes impact not only on the global north but also the south. It relates these debates to the increasing significance of rentier capitalism and the articulation of class, race and gender. The following chapter directs our attention towards debates concerned with post-capitalism, post-work and is one that is lodged within an argument that seeks to move beyond narrow conceptualisations of waged labour and by default VET. We need to think about the manner in which VET could conjoin with a rights-based understanding of education. Such a model could be moulded to address the needs and interests of those placed outside formal waged relations involved in various forms of unwaged labour that serve their communities and contribute to well-being.

Acknowledgement This chapter draws in part on my 'Socio-technical imaginary of the fourth industrial revolution and its implications for vocational education and training: a literature review', published in the Journal of Vocational Education & Training, 2018. 70:3, 337–63, https://doi.org/10.1080/13636820.2018.1498907

Note

1. The Pew Research Center describes itself as 'a nonpartisan fact tank that informs the public about the issues, attitudes and trends shaping the world'. (https://www.pewresearch.org/about/)

References

Acemoglu, D., and P. Restrepo. 2017. Robots and Jobs: Evidence from US Labor Markets. NBER Working Paper No. 23285. National Bureau of Economic Research. http://www.nber.org/papers/w23285.pdf

Adler, P. 2006. From Labor Process to Activity Theory. In *Critical Perspectives on Activity: Explorations across Education, Work and Everyday Life*, ed. P.H. Sawchuk, N. Duarte, and M. Elhammoumi, 225–261. Cambridge: Cambridge University Press.

Adler, P., and C. Heckscher. 2006. Towards Collaborative Community. In *The Firm as a Collaborative Community*, ed. C. Heckscher and P. Adler, 11–105. Oxford: Oxford University Press.

Allman, P., P. McLaren, and G. Rikowski. 2003. After the Box People. In *Yesterday's Dreams: International and Critical Perspectives on Education and Social Class*, ed. J. Freeman-Moir and A. Scott, 149–179. Christchurch: Canterbury University Press.

Anderson, D. 2009. Productivism and Ecologism: Changing Dis/courses in TVET. In *Work, Learning and Sustainable Development*, ed. J. Fien, R. Maclean, and Man-Gon Park, 35–57. Dordrecht: Springer.

Anslow, L. 2016. Robots Have Been about to Take All the Jobs for More than 200 Years: Is It Really Different This Time? https://timeline.com/robots-have-been-about-to-take-all-thejobs-for-more-than-200-years-5c9c08a2f41d

Atkinson, R., and J. Wu. 2017. False Alarmism: Technological Disruption and the U.S. Labor Market, 1850–2015, Information Technology and Innovation Foundation. http://www2.itif.org/2017-false-alarmism-technological-disruption.pdf?_ga=2.144641148.812496487.1496450124-341239608.1496450124

Autor, D. 2013. The Task Approach' to Labor Markets: An Overview. *Journal for Labour Market Research* 46(3): 185–199.

———. 2015. Why Are There Still so Many Jobs? The History and Future of Workplace Automation. *The Journal of Economic Perspectives* 29 (3): 3–30. https://doi.org/10.1257/jep.29.3.3.

Autor, D., F. Levy, and R. Murnane. 2003. The Skill Content of Recent Technological Change: An Empirical Exploration. *The Quarterly Journal of Economics* 118 (4): 1279–1334. https://doi.org/10.1162/003355303322552801.

Autor, D., L. Katz, and M. Kearney. 2006. The Polarization of the U.S. Labor Market, the American. *Economic Review* 96 (2): 189–194. https://doi.org/10.1257/000282806777212620.

Avent, R. 2017. *The Wealth of Humans*. London: Penguin.

Avis, J. 2010. Workplace Learning, Knowledge, Practice and Transformation. *Journal for Critical Education Policy Studies* 8 (2): 165–193. http://www.jceps.com/?pageID=article&articleID=197.

———. 2016. *Social Justice, Transformation and Knowledge: Policy. Workplace Learning and Skills*. London: Routledge.

———. 2018. Crossing Boundaries: VET, the Labour Market and Social Justice. *International Journal for Research in Vocational Education and Training* 5 (3): 178–190. https://doi.org/10.13152/IJRVET.5.3.2.

———. 2020a. Thinking About the Future: The Fourth Industrial Revolution, Capitalism, Waged Labour and Anti-work. In *The SAGE Handbook of Learning and Work*, ed. M. Malloch, L. Cairns, K. Evans, and B. O'Connor. London: Sage (Under review).

———. 2020b. Whither a Politics of Hope: Neoliberalism and Revolutionary Reformism? In *Caliban's Dance*, ed. M. Daley, K. Orr, and J. Petrie, 7–13. London: UCL IOE Press.

Ball, S., M. Maguire, and A. Braun. 2011. *How Schools Do Policy: Policy Enactments in Secondary Schools*. London: Routledge.

Bastani, A. 2019. *Fully Automated Luxury Communism*. London: Verso.

Baxter, G., J. Rooksby, Y. Wang, and A. Khajeh-Hosseini. 2012. The Ironies of Automation … Still Going Strong at 30? Proceedings of ECCE 2012 Conference, 29th–31st August, Edinburgh, North Britain, 65–71. https://doi.org/10.1177/1753193411414639

Bernstein, B. 1977. *Class, Codes and Control*. Vol. 3. London: RKP.

Bodrožić, Z., and P. Adler. 2018. The Evolution of Management Models: A Neo-Schumpeterian Theory. *Administrative Science Quarterly* 63 (1): 85–129. https://doi.org/10.1177/0001839217704811.

Boutang, Y.M. 2011. *Cognitive Capitalism*. Cambridge: Polity.

Braverman, H. 1974. *Labor and Monopoly Capital: The Degradation of Work in the Twentieth Century*. New York: Monthly Review Press.

Bregman, R. 2016. *Utopia for Realists*. Amsterdam. the Correspondent. https://thecorrespondent.com/.

Brown, P., H. Lauder, and D. Ashton. 2011. *The Global Auction*. Oxford: Oxford University Press.

Brynjolfsson, E., and A. McAfee. 2011. *Race Against the Machine: How the Digital Revolution Is Accelerating Innovation, Driving Productivity, and Irreversibly Transforming Employment and the Economy*. Lexington, MA: Digital Frontier Press.

———. 2014. *The Second Machine Age: Work, Progress and Prosperity in a Time of Brilliant Technologies*. New York: Norton & Company.

Chollet, F. 2017. The Impossibility of Intelligence Explosion, Medium. https://medium.com/@francois.chollet/the-impossibility-of-intelligence-explosion-5be4a9eda6ec.

Cirillo, V. 2018. Job Polarization in European Industries. *International Labour Review* 157 (1): 39–63.

Cleaver, H. 2017. *Rupturing the Dialectic*. Chicago: AK Press.

Coyle, D. 2017. Precarious and Productive Work in the Digital Economy. *National Institute Economic Review* 240: 5–14. https://doi.org/10.1177/002795011724000110.

Dittrich, P. (2016) Reskilling for the Fourth Industrial Revolution: Formulating a European Strategy. Jacques Delors Institut – Berlin. http://www.institutdelors.eu/wp-content/uploads/2018/01/digitalskill-jdib-nov2016.pdf?pdf=ok.

DWP. 2019. Employment Rate Remains at Record High. https://www.gov.uk/government/news/employment-rate-remains-at-record-high

Dyer-Witheford, N., A.M. Kjøsen, and J. Steinhoff. 2019. *Inhuman Power*. London: Pluto.

———. 2010. *From Teams to Knots: Activity-Theoretical Studies of Collaboration and Learning at Work*. Cambridge: Cambridge University Press.

Fleming, P. 2017. *The Death of Homo Economicus*. London: Pluto.

Ford, M. 2016. *The Rise of the Robots*. London: Oneworld.

Fraser, N., and R. Jaeggi. 2018. *Capitalism: A Conversation in Critical Theory*. Cambridge: Polity.

Frey, C. B., and M. A. Osborne. 2013. The Future of Employment: How Susceptible are Jobs to Computerisation? Published by the Oxford Martin Programme on Technology and Employment. https://www.oxfordmartin.ox.ac.uk/publications/view/1314

Goos, M., and A. Manning. 2007. Lousy and Lovely Jobs: The Rising Polarization of Work in Britain. *The Review of Economics and Statistics* 89 (1): 118–133. https://doi.org/10.1162/rest.89.1.118.

Goos, M., A. Manning, and A. Salomons. 2014. Explaining Job Polarization. *The American Economic Review* 104 (8): 2509–2526. https://doi.org/10.1257/aer.104.8.2509.

Gorz, A. 2010. *The Immaterial*. London: Seagull.
Guile, D., and L. Unwin. 2019a. Introduction to the Handbook: Vocational Education and training (VET) Theory, Practice, and Policy for a Complex Field of Inquiry. In *The Wiley Handbook of Vocational Education and Training*, ed. D. Guile and L. Unwin, 1–16. London: Wiley Blackwell.
———. 2019b. VET, Expertise, and Work: Situating the Challenge for the Twenty-first Century in Guile. In *The Wiley Handbook of Vocational Education and Training*, ed. D. Unwin and L. Unwin, 19–40. London: Wiley Blackwell.
Haldane, A. 2015. Labour's Share Speech Given by Chief Economist, Bank of England. https://www.bankofengland.co.uk/-/media/boe/files/speech/2015/labours-share.pdf?la=en&hash=D6F1A4C489DA855C8512FC41C02E014F8D683953
Harvey, D. 2014. *Seventeen Contradictions and the End of Capitalism*. London: Profile.
Hodkinson, P., and D. James. 2003. Transforming learning cultures in further education. *Journal of Vocational Education and Training* 55 (4): 389–406. https://doi.org/10.1080/13636820300200236.
Holloway, J. 2019. *We are the Crisis of Capitalism Oaklands*. Kairos PM.
Holzer, H. 2015. Job Market Polarization and U.S. Worker Skills: A Tale of Two Middles. Brookings Institution Economic Studies Working Paper. https://www.brookings.edu/wp-content/uploads/2016/06/polarization_jobs_policy_holzer.pdf
Ireland, C. 2018. *Reducing Public Speaking Anxiety in Undergraduates: a Case Study of an Intervention with Accountancy Students*. Doctoral Thesis, University Of Huddersfield.
Keep, E., and S. James. 2012. A Bermuda Triangle of Policy? 'Bad Jobs', Skills Policy and Incentives to Learn at the Bottom End of the Labour Market. *Journal of Education Policy* 27 (2): 211–230.
Keynes, J.M. 2009. Economic Possibilities for Our Grandchildren (1930). In *Essays in Persuasion*, 191–202. New York: Classic House books.
Livesey, F. 2017. *From Global to Local*. London: Profile books.
Livingston, J. 2016. *No More Work*. Chapel Hill: University of North Carolina Press.
Lotringer, S. 2004. Foreword. In *A Grammar of the Multitude*, ed. P. Virno. Los Angeles: Semiotext(e).
Martin, R., and P.S. Morrison. 2003. The Geographies of Labour Market Inequality. In *Geographies of Labour Market Inequality*, ed. R. Martin and P.S. Morrison. London: Routledge.

Marx, K. 1976 [1867]. *Capital: A Critique of Political Economy*, Vol 1. Harmondsworth: Penguin

Mason, P. 2015. *Postcapitalism*. London: Allen Lane.

Mason, P. 2019. *Clear Bright Future*. London: Allen lane.

McAfee, A. 2019. *More from Less*. London: Simon and Schuster.

McGrath, S., and L. Powell. 2016. Skills for Sustainable Development: Transforming Vocational Education and Training Beyond 2015. *International Journal of Educational Development* 50 (2016): 12–19.

McRae, H. 2019. The Job Market is Booming—but Not for Obvious Reasons. *The Independent* (Daily Edition), September 11, p. 35.

MGI. 2013. Disruptive Technologies: Advances that Will Transform Life, Business, and the Global Economy. Tech. Rep., McKinsey Global Institute. https://www.mckinsey.com/business-functions/digital-mckinsey/our-insights/disruptive-technologies

Mishel, L., and J. Bivens. 2017. The Zombie Robot Argument Lurches On, Economic Policy Institute. http://www.epi.org/publication/the-zombie-robot-argument-lurches-on-there-isno-evidence-that-automation-leads-to-joblessness-or-inequality/?utm_content=buffereba76&utm_medium=social&utm_source=twitter.com&utm_campaign=buffer

Mishel, L., and H. Shierholz. 2017. Robots, or Automation, are Not the Problem: Too Little Worker Power Is. Economic Policy Institute. https://www.epi.org/publication/robots-orautomation-are-not-the-problem-too-little-worker-power-is/

Mokyr, J., C. Vickers, and N. Ziebarth. 2015. The History of Technological Anxiety and the Future of Economic Growth: Is This Time Different? *Journal of Economic Perspectives* 29 (3): 31–50. https://doi.org/10.1257/jep.29.3.31.

Mould, O. 2018. *Against Creativity*. London: Verso.

Nevin, S. 2016. Lifelong Learning Reforming Education for an Age of Technological and Demographic Change. IoD policy report. https://www.iod.com/Portals/0/PDFs/Campaigns%20and%20Reports/Employment%20and%20Skills/Life%20Long%20Learning%20Report.pdf?ver=2016-09-14-124014-230.

Orr, K. 2016. *Who Cares About Vocational Education and Training? Pedagogy and Pathways for the 'overlooked middle'*. Professorial inaugural lecture, The University of Huddersfield, December 1.

Paulsen, R. 2014. *Empty Labor: Idleness And Workplace Resistance*. Cambridge: Cambridge University Press.

Peters, M.A. 2017. Technological Unemployment: Educating for the Fourth Industrial Revolution. *Educational Philosophy and Theory* 49 (1): 1–6. https://doi.org/10.1080/00131857.2016.1177412.

Pew Research Center. 2014. AI, Robotics, and the Future of Jobs. https://www.pewinternet.org/2014/08/06/future-of-jobs/

Pfeiffer, S. 2014. Digital Labour and the Use-Value of Human Work. On the Importance of Labouring Capacity for Understanding Digital Capitalism. *tripleC* 12 (2): 599–619. https://doi.org/10.31269/triplec.v12i2.545.

———. 2016. Robots, Industry 4.0 And Humans, or Why Assembly Work Is More than Routine Work, Societies. *Societies*, June. 1–26. https://doi.org/10.3390/soc6020016

Rifkin, J. 1995. *The End of Work*. New York: Putnam.

———. 2014. *The Zero Marginal Cost Society*. London: Palgrave Macmillan.

Roberts, K. 2016. Youth Cultures and the Formation of a New Political Generation in Eastern Europe. In *Eastern European Youth Cultures in a Global Context*, ed. M. Schwartz and H. Winkel, 46–63. London: Palgrave Macmillan.

Rodrik, D. 2016. Premature Deindustrialization. *Journal of Economic Growth* 2 (1): 1–33. https://doi.org/10.1007/s10887-015-9122-3.

Ross, A. 2016. *The Industries of the Future*. London: Simon &Schuster.

Sale, K. 1996. Rebels against the Future: The Luddites and Their War on the Industrial Revolution. Boston: Addison Wesley.

Saniter, A., and F. Howe. 2017. Industry 4.0—What's behind the Mask? A Case of Addititve Manufacturing. In *Social Dimensions and Participation in Vocational Education*, ed. F. Kaiser and S. Krugmann, 172–176. Rostock: University of Rostock.

Seeverband der Bayerischen Metall- und Elektro-Arbeitsgeber. 2016. Industrie 4.0–Auswirkungen Auf Asund Weiterbildung in Der M+E Industrie. https://www.vme-net.de/sites/default/files/downloads_und_vorschaubilder/baymevbm_studie_industrie-4-0.pdf.

Sennett, R. 2009. *The Craftsman*. Harmondsworth: Penguin.

Shildrick, T., R. MacDonald, C. Webster, and K. Garthwaite. 2012. *Poverty and Insecurity: Life in low-pay, No-pay Britain*. Bristol: Policy Press.

Solow, R. 2008. The German Story. In *Low-wage Work in Germany*, ed. G. Bosch and C. Weinkopf, 1–14. New York: The Russell Sage Foundation.

Streeck, W. 2016. *How Will Capitalism End?* London: Verso.

Susskind, R., and D. Susskind. 2016. *The Future of the Professions*. Oxford: Oxford University Press.

Teixeira, R. 2017. It's Time to Explode the Myth of "McJobs". https://www.vox.com/the-bigidea/2017/6/13/15788610/robots-artificial-intelligence-ai-mcjobs-low-skill.

Victor, B., and A. Boynton. 1998. *Invented Here: Maximizing Your Organization's Internal Growth and Profitability*. Harvard: Harvard Business School Press.

Wallerstein, I., R. Collins, M. Mann, G. Derluguian, and C. Calhoun. 2013. *Does Capitalism Have Future?* Oxford: Oxford University Press.

Wheelahan, L. 2019. Knowledge, Competence, and Vocational Education. In *The Wiley Handbook of Vocational Education and Training*, ed. D. Guile and L. Unwin, 97–112. London: Wiley.

Whittaker, M. 2015. What Do Current Trends Tell Us about the British Labour Market of Tomorrow? In *Technology, Globalisation and the Future of Work in Europe: Essays on Employment in A Digitised Economy*, ed. T. Dolphin, 36–41. London: IPPR. https://www.ippr.org/publications/technology-globalisation-and-the-future-of-work-in-europe.

Winch, C. 2012. *Dimensions of Expertise*. London: Continuum.

Wolf, M. 2019. Why Rigged Capitalism is Damaging Liberal Democracy. *Financial Review*, September 19. https://www.afr.com/policy/economy/why-rigged-capitalism-is-damaging-liberal-democracy-20190919-p52su6

Young, M. 2008. *Bringing Knowledge Back In*. London: Routledge.

4

Post-Work, Post-Capitalism and the Fourth Industrial Revolution

Abstract The chapter addresses arguments about Italian Workerism, post-capitalism, post-work, universal basic income and the politics of abundance. It moves beyond current conceptualisations of the fourth Industrial Revolution (4th IR) to develop a politically informed analysis that can inform a critical, if not transformative praxis. Labour market analysts refuse the suggestion that labour is being expelled from paid employment. There is a tension between those who experience and bemoan precariousness, intermittent and under-employment and those for whom all of life is centred around work. For some this had led to their expulsion from waged labour, whilst for others it has led to its intensification. The 4th IR is presented as inevitable, silencing the manner in which it relates to the development of capitalism and its articulation with class struggle. The chapter calls for an expansive understanding of both class and vocational education and training.

Keywords Fourth Industrial Revolution • Post-work • Post-Capitalism • Universal Basic Income • Workerism

J. Avis, *Vocational Education in the Fourth Industrial Revolution*,
https://doi.org/10.1007/978-3-030-52032-8_4

Introduction

This chapter draws on arguments raised in previous chapters that address the future of work and skill, setting this alongside more radical analyses. Much of the work that addresses the labour market refuses the suggestion that labour is being expelled from paid employment and stands in direct opposition to some of the analyses considered earlier in the book. There is a tension between the experiences of those who bemoan precariousness, intermittent and under-employment and those for whom the whole of life seems to be centred around work. Paradoxically, for some this had led to the expulsion from waged labour, whilst for others it has led to its intensification. In its ideological manifestation the fourth Industrial Revolution (4th IR) is presented as an inevitable development flowing from the first to the current IR. However, what is silenced in these constructions is the manner in which the 4th IR connects to the development of capitalism and its articulation with class struggle. These processes alert us to the way in which technology is entwined with social relations. Consequently, this chapter addresses arguments that engage with Italian Workerism, post-capitalism, post-work, universal basic income (UBI) and, paradoxically, the politics of abundance. It seeks to develop a politically informed analysis that moves beyond current conceptualisations of the 4th IR that are lodged within productivism, which in turn can inform a critical, if not transformative praxis (Anderson 2009; McGrath and Powell 2016; Powell 2012).

The Labour Market

Frey (2019) argues that technology can impact upon the labour market and waged labour in two ways. It can lead to the displacement of labour, technological unemployment or alternatively can augment the labour process resulting in increased productivity and opportunities for employment. In addition, on the basis of an engagement with a wide-ranging literature, Frey argues that there are similarities between the disruptions that ensued following the 1st IR and current conditions which have been described as the 4th IR (see Fig. 1.1, Chap. 1). In the early years of the 1st IR following mechanisation, workers were displaced from employment leading to a fall in the wages of formerly skilled workers, as did the ranks of those they joined, semi- and unskilled workers. 'During the first four decades of the nineteenth century, the profit share of national income doubled, as both the share of land and labor declined' (Frey 2019: 132). Drawing upon Allen (2009), Frey describes this period as 'Engels' pause'. Marx and Engels analysis was set within this particular moment of capitalist development with the Marxist critique emphasising the immiseration of the proletariat. For both Allen and Frey this period was followed by the second IR which was characterised by the introduction of technology which served to augment labour power and consequently resulted in increased wages for the working class—that is to say those who laboured in semi- and unskilled work. In an echo of the embourgeoisement thesis these workers were able to attain the life style and material well-being of the middle class. There are at least two points to note. It took something like 40 years for the disruption following the 1st IR to be resolved in favour of the working class with an expansion of the need for semi- and unskilled labour. During this period technology augmented labour enabling workers to be more productive and therefore empowered to enhance not only their wages but also material and social well-being. Whilst Frey is rightly cautious about speculating on the future of work, his argument suggests that we are currently in a conjuncture in which technology is displacing the labour of semi- and unskilled workers in much the same way as occurred during 'Engels' pause'. At the same time technology has spawned the need for high skilled, highly educated labour, resulting in an increasing divide between the university and highly educated middle class and the working class. Yet the logic of the argument would seem to suggest that if parallels can be drawn between the 1st and 4th IR, with the 2nd resolving the dislocation caused by the 1st, it would seem likely that after a period of disruption there would be a renewed need for semi- and unskilled labour (and see Autor 2015). That is to say, that in the same way as the 2nd IR sought to augment the labour power of semi- and unskilled workers the future may hold not dissimilar possibilities for those workers who are currently displaced. If this argument is accepted the Keynesian and social democratic settlement that occurred in the twentieth century should not be considered as exceptional but rather as an aspect of capitalist development. Frey would argue that such a future is far from certain. However, what he would unequivocally claim is that the current moment is one in which technology is replacing the labour of the semi- and unskilled working class with all the negative consequences that flow from that.

Whatever the future holds for waged labour, it will in part be shaped by political and social processes that coalesce around an expansive notion of class struggle as well as by the affordances offered by technology, be it enabling or displacing workers. It is also important that we acknowledge that technology comes from somewhere and is never innocent or neutral. It derives from particular socio-economic circumstances and the balance of power between labour and capital, seeking to address particular problems and issues. An expansive notion of class struggle will incorporate intersectionality not only of the workless and unwaged but also members of the middle class (Avis 2019). It also calls for a re-imagining of what constitutes vocational education and training (VET) in conditions of worklessness, the salience of unwaged labour and the significance of the informal economy, particularly in the social formations of the global south (McGrath and Powell 2016; Powell 2012).

Whilst the spectre of technological unemployment may appear to be exaggerated, it is vital to consider the type of employment that is available to displaced workers, in terms of skill, duration and remuneration. I am somewhat uneasy with the use of the terms semi- and unskilled work. Whilst it is the undoubtedly the case that the formerly well-paid factory workers of the Keynesian era have been displaced, and more often than not find themselves in what are considered low-skilled and low-waged routine jobs as janitors, care and service workers and so on, nevertheless we need to interrogate the way skill is understood in these contexts. One way to do this is to think about the skills of the professional middle class and what we might describe as really useful labour, or indeed that which is productive, that is to say, embodying use value. This raises a range of questions about the way in which we understand these terms which I address below.

Thinking About Skill

Earlier in this book, I examined the relationship between skill, expertise and by default connoisseurship. I argued that skill is not so much a feature of the individual but rather a collective accomplishment and it is for this reason that expertise would be a better term as it embodies a developmental and dynamic trajectory. Not only is expertise continuously developed it is also situational, contingent and will involve the exercise of judgement and connoisseurship that develops over time whereby the practitioner gains a deep and almost intuitive understanding of the field of practice. However, the enactment of expertise will nevertheless be constrained not only by the immediate but also by the broader socio-economic context in which it is placed.

In recent years much has been made of the skills required by the 'knowledge' or digital economy. The World Economic Forum (WEF 2018) in its report *The Future of Jobs* surveyed senior executives to ascertain the top ten key skills currently required and those they anticipated would be in demand in 2022. The main point to be made is that alongside the proficiency in new technologies it was anticipated that there would be a significant demand for soft, human and cognitive skills.

2018	2022
Analytical thinking and innovation	Analytical thinking and innovation
Complex problem-solving	Active learning and learning strategies
Critical thinking and analysis	Creativity, originality and initiative
Active learning and learning strategies	Technology design and programming
Creativity, originality and initiative	Critical thinking and analysis
Attention to detail, trustworthiness	Complex problem-solving
Emotional intelligence	Leadership and social influence
Reasoning, problem-solving and ideation	Emotional intelligence
Leadership and social influence	Reasoning, problem-solving and ideation
Co-ordination and time management	Systems analysis and evaluation

(Derived from WEF 2018: 12)

Interestingly and perhaps unsurprisingly these key skills articulate with those which are putatively associated with the professional middle class.

Amongst these skills are those that involve a commitment to lifelong learning, analytic thinking, creativity, critical thinking, problem-solving, the use of emotional intelligence and leadership skills that can be used to secure influence. There is a clear echo not only with Reich's (1992) 'symbolic analyst' but also with the flexible, adaptive, enterprising subject of neo-liberalism who constantly refreshes their skills—the high flying graduate employee. These highly skilled individuals will constitute the type of knowledge workers that Brown et al. (2011: 81) describe as developers, top performers who are given 'permission to think'. Importantly these knowledge workers are placed alongside others, who Brown et al. (2011: 81) describe as demonstrators and drones, terms which themselves reflect the de-skilling and routinisation of knowledge work. This labour lends itself to the mobilisation of artificial intelligence (AI) which results in the hollowing out of middle-level knowledge work.

There are several additional arguments to be addressed not least amongst which is the way in which those with social and economic power use this influence to attribute to themselves rare and remarkable skills which sets them apart from the rest of us (see e.g. Davis and Moore 1945). Both historically and contemporaneously this over-valuation of particular forms of labour and skill has rested alongside the devaluation of other types of service work. This devaluation has frequently been applied to the 'care' work carried out by women as well as other marginalised groups, ethnic minorities, immigrants and so on in wider society. Care work because of its complexity and demand for emotional intelligence is difficult to automate and requires a number of the key skills that the World Economic Forum suggests are not only essential in 2018 but likely to be needed in 2022. Many business leaders consider these skills are expected to emerge in the near future and, in the WEF's terms, are 'trending'. We could think about analytic and critical thinking, active learning, creativity, problem-solving, emotional intelligence and reasoning—all of these skills could be mobilised by a care worker attempting to administer to the complex and demanding needs of an elderly person suffering from dementia. We could think of this as 'really useful labour' as it contributes to the well-being of those it caters for and therefore has 'real' use value. The importance of the nursing and social care workforce has been strongly evidenced during the COVID-19 pandemic. However, it should not be overlooked that care work has been commodified and is subject to the disciplines of the capitalist labour process whereby its aim is to generate surplus value rather than to meet the needs of those it serves. In this way care work could be thought of as productive labour in the sense that it serves to generate profit. To the extent that care work contributes to well-being this is frequently in spite of the capitalist relations in which it is set and is undertaken at some cost, both emotionally and economically, to the care worker. Thus in contradiction to the previous point this labour nevertheless could be considered productive in the sense that it develops use value and well-being.

The evaluation of work is in part dependent on who labours and the social location in which this arises which in turn constitutes a value judgement. Jeremy Corbyn, the then leader of the Labour party, in his speech launching the 2019 election campaign stated:

> And the big question of this election is: whose side are you on? Are you on the side of the tax dodgers … The dodgy landlords … The bad bosses … The big polluters … The greedy bankers … The billionaire media barons …, whose empire pumps out propaganda to support a rigged system.
>
> You know whose side Labour's on—a Labour government will be on your side. Together, we can pull down a corrupt system and build a fairer country that cares for all.

Interestingly Corbyn's invective was directed at the corruptions of the capitalist system. This is a theme that is present in much of the writing of those who support capitalism but who bemoan its excesses (Rajan 2019; Stiglitz 2019) and who anticipate the development of a fairer and more caring form of capitalism. There is an echo here of social democracy and Nordic egalitarian capitalism (Kenworth 2004). However, the logic of capitalism is that it will endlessly seek to accumulate capital and develop new markets, as can be seen in the privatisation of parts of the welfare state (Harvey 2014: 23–24). At times when labour is in ascendance and wields greater power the excesses of capitalism will be constrained through class struggle. However, at other times such as in recent decades in which neo-liberalism has gained ascendancy capital will pursue its goals aggressively. In the current conjuncture a number of the key skills that the WEF cites will be mobilised by members of the professional middle class in the service of capital and could thereby be seen as a form of productive labour, in that it serves to facilitate the accumulation of capital. Yet at the same time as Graeber (2018) argues it is broadly within this professional constituency that we encounter what he describes as 'bullshit jobs', ultimately meaningless and in any definition of work, unproductive labour. Aspects of the academic labour process would be a case in point reflected in the time spent in committee meetings, dealing with emails and so on (Fleming 2017: 137; Zukas and Malcolm 2015: 7–11). Graeber (2018: 217) writes,

> Some academic environments are more anti-intellectual than others. But everywhere, at the very least, there is a sense that the pleasurable aspect of one's calling, such as thinking, were not really what one is being paid for; they were better seen as occasional indulgencies one is granted in recognition of one's real work which is largely about filling in forms.

Graeber is criticised for drawing on personalised accounts that do not go beyond the subjective and thereby fails to consider their relationship to capitalist and neo-liberal processes. He does nevertheless draw our attention to the messiness of these processes and the manner in which they articulate to the interests of the powerful. For example, the preoccupation of many organisations with performativity and the measurement of outcomes aligns with low trust relations whereby professionals are deemed untrustworthy and therefore need to be constantly monitored. In addition, meaningless labour may be used to buttress the power and prestige of organisations and senior management teams. In these instances, Graeber (2018: 13) cites the work of receptionists, front of house workers in office blocks who have little to do, but whose presence reaffirms the power and prestige of the organisation. Relatedly Anderson (2017) refers to the dictatorship of 'private governance' whereby employees are subject to the unaccountable whims of senior management. There are at least two points to be made. First, the arbitrary rule of management cadres is subject to the disciplines of capitalism this can be softened by monopolistic and oligarchic tendencies which serve to concentrate power and facilitate its abuse. Second, the increasing significance of 'financialisation' has interrupted the 'traditional' way in which surplus value is appropriate, that is money being used to generate more money (M-M), thereby circumventing the role of productive labour. This process has rather more in common with primitive accumulation and rent seeking than with the real subsumption of labour to capital. Notably, the current conjuncture is one which features the decoupling of employment and wages from productivity and growth. This can be seen in the arbitrariness of the relationship between wages and productivity; we need only to reflect on the excessive incomes awarded to senior management whose companies fail, setting this against the derisory wages of productive workers. At times this can be reflected in the irrationality and incoherence surrounding the distribution of income and wealth. This distribution is of course shaped by the balance of class power which currently favours senior management and the interests of capital.

Species Being

Waged labour is central to the current doxa and is construed as being pivotal to social and economic well-being with its absence seen as the obverse. The lack of waged labour carries with it a range of social harms that can lead to a sense of worthlessness resulting in both physical and mental health issues. Work is construed as being central to who and what we are (Wilkinson and Pickett 2010, 2018). This raises several issues. The significance of waged labour is that it can enable full participation in wider society. However, this is predicated on having a sufficient income—notably in-work poverty is increasingly a feature of capitalist economies (Anderson 2017: 134–135; Bloodworth 2018; Ehrenreich 2010). In-work poverty is also reflected in highly alienating and unpleasant work in both the global north and south (The Guardian 2015; Parker and Vonow 2017; War on Want 2019a, b). Whilst what constitutes alienating and unpleasant work is not only a subjective but also a value judgement, we should nevertheless reflect on the type of work available to the working-class communities that Shildrick et al. (2012) studied. In these communities many people churned between low-waged, low-skilled work interspersed with periods of unemployment, but nevertheless the search for and commitment to waged labour is a feature of these communities. For a number of writers this commitment is used to challenge 'pathological' notions which construe such communities as not only refusing to work but as also living off state benefits and petty crime (Shildrick et al. 2012; Simmons and Thompson 2011). In this instance, the commitment to and search for decent work is seen as a social justice question. However, there is an irony. On one level this argument reflects a realistic political strategy that seeks to provide decent work for those who have been excluded. In some respects the debate concerned with UBI addresses these social justice questions. It is suggested that UBI would enable people to avoid highly exploitative and oppressive waged labour. It might lead employers to invest in capital machinery so as to remove the drudgery from waged labour whilst simultaneously encouraging improvements in the quality of work. At best UBI could be seen as an aspect of revolutionary reformism—a stage along the route to a qualitatively transformed social and economic system. Yet on another level through the centrality accorded to waged labour UBI may serve to lock people into capitalist work relations, becoming a form of entrapment (Holloway 2019: 230). For Cleaver (2017: 13) the struggle against waged labour, or in workerist terms its refusal, places 'the struggle against work [as] the most fundamental threat to the [capitalist] machine'. It is important to acknowledge the damage waged labour inflicts upon those who encounter in-work poverty. At the same time we also need to recognise the harm that results from the intensification of labour and the surrounding insecurities that arise when all of life revolves around work (Beckett 2018, online; Cederström and Fleming 2012; Frayne 2019; Marazzi 2011: 113). Sage (2019) points out that it is not so much the lack of the experience of work that matters but rather the failure to adhere to a powerful social norm that calls for involvement in waged labour. He claims that those who are weakly attached to this social norm suffer far less from the absence of waged labour than those who are strongly attached (Frayne 2015). For example, Sage (2019: 210) cites Hetschko et al. (2014: 150) who found that the well-being of those who transition from long-term unemployment to retirement increased significantly suggesting this was because they are no longer subject to the 'work ethic' and the opprobrium of not having waged employment (and see Ponomarenko et al. 2019).

Before considering the debates that surround the UBI, I would like to revisit conceptualisations of work. Much of the previous discussion prioritises waged labour and presupposes the patterns of employment that characterised industrial capitalism. It bemoans the transitions that have occurred in recent years that have led to exclusion from waged employment and increased precariousness. Malloch et al. (2021) argue that historically we have tended to define ourselves in terms of waged employment (see Wilkinson and Pickett 2010, 2018). Consequently, lifelong learning has been justified through its instrumental connection with waged work, this being particularly the case with VET. In this instance, Sage's (2019) argument is particularly salient, pointing as it does to those who have transitioned from waged employment to retirement. This argument serves to alert us to the potential contribution that a re-imagined VET could make to lifelong learning for those at different stages in their 'lifespace' and 'lifespan' and who are differently positioned to waged work (Malloch et al. 2021). Such a stance would call for the re-imagining of VET that moves beyond a narrow instrumentalism tied to waged employment.

Universal Basic Income

There are a range of conflicting arguments derived from both the left and right that revolve around UBI and its cognates (see Martinelli 2017, for a measured account of the issues as they apply to the UK). Much of this debate concerns questions of affordability and is located within a policy science framework that addresses the application of UBI in the current socio-economic context. This takes for granted existing capitalist relations and seeks to insert UBI into the current welfare system, one increasingly seen as out of kilter with the nature of waged labour and family life in the twenty-first century. That is to say, one in which Fordist models of employment have been superseded by more flexible and less secure work, with the two income family becoming normalised. Demographic changes in global north have led to increasing number of elderly people living in single occupancy households who require support by carers or family members. It is important to recognise that the manner in which UBI could be implemented varies with the specific version that is adopted. Yet at the same time within this policy science framework the lineaments of the argument remain the same being split between two diametrically opposed positions. Martinelli (2017: 2) suggests that 'rather than claiming UBI is unaffordable *per se*, a more apt characterisation of opposition is that *an affordable UBI would be inadequate, and an adequate UBI would be unaffordable*'. In addition, there are claims that UBI would not necessarily benefit the poorest members of society, that is, if UBI served to replace both the bureaucracy and range of benefits provided by the welfare state. The suggestion is that if UBI is to be affordable it would need to be set at such a low level that it would fail to address the poverty of the poorest and that if it was set at a sufficiently high level it would be unaffordable (Hoynes and Rothstein 2019). It is also suggested that UBI by definition would be a universal benefit offered to all and therefore its distribution would sit alongside existing patterns of inequality unless elements of conditionality were applied which would mean it would be bought into alignment with other welfare benefits, with all the costs that this would incur. Paradoxically, its effectiveness as a counter to poverty would be undermined by its universality which would mirror existing patterns of inequality (Goulden 2018).

For those who argue in favour of UBI from a policy science or technicist framework, it is thought that a basic income will address the inequalities that derive from precarious work as well as those forms of employment that lead to in-work poverty. A basic income would guarantee a minimum standard of living that would compensate for periods of unemployment as well as for inadequate wages (Skidelsky 2016). In addition, by reducing levels of poverty UBI would also bring with it improved physical and mental health (Haagh and Rohregger 2019) (World Health Organisation 2019). It is suggested a basic income would reduce welfare bureaucracy by shrinking the number of means tested benefits. In its pure form it would result in a universal payment to all members of society, placing on recipients the responsibility for how it was spent and the purposes to which it was put. In this way it would reduce the cost of the welfare state by eradicating all other benefits. However, many of the proposed versions that seek to implement UBI are based on some sort of conditionality. For example, the Royal Society of Arts' (RSA's) version reduces UBI once incomes are in excess of £75,000 pa and tapers it away to nothing for those who earn above £150,000 pa (Painter and Thoung 2015: 23).

In the following I want to move away from policy science, or what I have described as technicist approaches to UBI. Such approaches are concerned with technical questions surrounding the implementation of UBI, its feasibility, as well as its costs and benefits. These technicist approaches would interrogate whether it is the best way to address poverty and inequality in the current socio-economic context. The starting point for this discussion is with the moral and ethical arguments that UBI could prefigure the development of a future society that moves beyond both waged labour and capitalist relations—a post-work, post-capitalist social formation. Weeks (2011: 12) reminds us that there are other ways in which work, or more precisely productive labour, could be organised than those found in capitalist economies.

In the preceding discussion it could be argued I have used the term productive labour somewhat promiscuously. The term itself is Janus-like in that it can point to the production of surplus value and its appropriation by capital but also to the production of use value that may lie outside the remit of capital, the domestic labour of women involved in social reproduction would be a case in point (Hester and Srnicek 2019). Rustin (2013: 13) reminds us that 'there is no conceivable material or technological excuse for unemployment, when there is abundant work which could and should be done, in nurturing, developing and expressing human capabilities'. Whilst this passage is rooted in productivism it nevertheless alerts us to the possibilities of 'really useful labour' lying outside of capitalist wage relations, in which we could express our species being. This would call for a form of VET predicated on the right to education which broke from its connection with productivism (Anderson 2009). Such a model would seek to enable participants to engage with really useful labour and would facilitate a re-•imagining of social and political relations, thereby allowing VET to be directly engaged with our species being.

Hunnicutt's (2013) discussion of the forgotten American dream is both salutary and pertinent. At the onset of industrial capitalism there was a concerted attempt to discipline workers so that they became habituated to the rhythms of the factory—immiseration was one such tactic. Hunnicutt draws our attention to the struggles of American workers in the 19th and the early years of the twentieth century for the reduction of working hours and their call for the exponential expansion of free time—the "progressive shortening of the hours of labor" (Hunnicutt 2013: xi). Such concerns were apparent until the middle years of the last century when they were eclipsed by the pursuit of an ever-increasing standard of living. Free time was a space in which workers could exercise their species being and engage effectively in un-alienated labour, thereby escaping the 'tyranny of the job' and reasserting the dignity of labour external to capitalist wage relations (Hunnicutt 2013: 7). Although the struggle against the discipline of the factory was evident in Britain during the early stages of industrialisation concerns with the extension of free time was less pronounced than in America. It was nevertheless a feature of trade union campaigns. However, this struggle was of a reformist kind, contributing towards the social reproduction of labour. It provided a space where workers could escape from the meaninglessness of waged labour. Engagement with free time enabled some workers to give meaning to their lives and facilitated their subsequent reengagement with paid employment. Current concerns with work-life balance resurrect such concerns. At the same time the call for the extension of free time could be reinvigorated to articulate with moves towards UBI as part of a revolutionary reformism. As Pitts and Dinerstein (2017a: 428) quizzically comment on this position 'basic income is perceived as the means by which the link between money and subsistence can be maintained in a world where the relationship with the wage weakens'. Yet, UBI may also place the capitalist state in a position of power over those whose income fails to meet their subsistence needs. At the same time UBI because of its focus on individualisation serves to undermine the collective power of labour and trade unions.

For writers, influenced by Italian Workerism, such as Fumagalli (2011), the relationship of labour and the measurement of surplus value has been interrupted in the current economic context. He suggests that 'with the advent of cognitive [knowledge, immaterial, info-] capitalism, valorization tends to graft itself onto different forms of labor, which go beyond the official work-time and coincide more and more with the whole lifetime' (2011: 10, and see Vercellone 2009). These arguments are buttressed by a particular analysis of the 'general intellect' and a passage in the Grundrisse referred to as 'fragments on machines' (Marx 1973 [1857–1858] 704–706, 2014 [1858]). Cognitive capitalism emphasises the role of 'common' collectively formed knowledge developed outside the labour process which is subsequently appropriated by capital in the pursuit of surplus value. Gorz (2010: 52) points out that cognitive capitalism operates in a different manner to industrial capitalism in that its main productive force, knowledge, is a product that is in large part the outcome of unpaid collective activity. In addition, the manner in which production is described in the 'fragments' is deemed to illustrate the logic of capitalist development. Marx (1973: 705) writes, 'Labour no longer appears so much to be included within the production process; rather, the human being comes to relate more as watchman and regulator to the production process itself'. The thrust of these arguments is to underpin those that celebrate the possibility of a post-capitalist, post-work society, one which is marked by abundance. This is allied with, what may be described as, following Adler (2007), paleo Marxism, whereby the full development of the forces of production are held in check by the relations of production. In these instances, knowledge itself is deemed to be an important means of production whose transfer and distribution incurs minimal costs leading to what Rifkin (2014) describes as 'the zero marginal cost society'. The full development of the forces of production heralds the possibility of a society marked by abundance in which traditional notions of waged labour are transcended. This 'accelerationist' stance is a feature of the work of Hester and Srnicek (2019), Mason (2015) and Srnicek and Williams (2015), amongst others. The political logic here is to push capitalism as far and as fast as possible, thereby exacerbating the contradiction between the forces and relations of production, speeding-up the move towards a post-work and post-capitalist social formation (Mackay and Avanessian 2014). Mason (2015: 144) writes,

> Technologically, we are headed for zero-priced goods, unmeasurable work, an exponential take off in productivity and the extensive automation of physical processes. Socially we are trapped in a world of monopolies, inefficiency, the ruins of a finance-dominated free market and a proliferation of 'bullshit jobs'.

For Mason this is reflected in the contradiction between 'the possibility of free abundant socially produced goods' and 'monopolies, banks and governments seeking to maintain control' (2015: 144). It is almost as if capitalism will bring about its own demise and that all we need to do is to work with the grain of the development of the forces of production to transform society.

Beyond Waged Labour

Whilst the move beyond waged labour is thought to be progressive and predicated on the development of the forces of production, this relationship is not straightforward. This issue is addressed in critiques of Workerism, post-work and post-capitalism which raise several questions concerned with: the analysis of general intellect, productive labour, forces of production as well as the measurement of value. Writers such as Pitts and Dyer-Witheford, amongst others, draw our attention to the difficulties and tensions in workerist, post-capitalist and post-work thinking (Dyer-Witheford et al. 2019; Pitts and Dinerstein 2017b). Pitts, for example, emphasises that the focus on waged labour fails to move beyond capitalist relations for two reasons: first, the provision of UBI replaces the employer with the capitalist state and, second, and more importantly for this discussion, analyses of post-work fail to transcend the social relations of production. Lombardozzi and Pitts (2019: 9) argue,

> What appears simply as a crisis of work, wage and social democracy in fact reflects a deeper and more pervasive crisis of social reproduction. These three are the social forms that mediate the antagonistic and contradictory struggle to survive in capitalist society. Combined with such a 'social form' perspective, social reproduction theory expands the theoretical understanding of the under-investigated intertwined relationship between productive and reproductive work which reproduces social inequality and hampers wellbeing.

Importantly for Pitts (2018: 3) the key insight of social form analysis is that value does not derive from concrete but rather from abstract labour which is 'a category of social mediation expressed in money', a social form undergirded by antagonistic social relations. It is here that the crisis of measurement can be resolved. This resolution is based on 'socially necessary labour-time, a retrospective abstract social relation between all labours mediated by commodity exchange' (Pitts 2018: 24).

Hester and Srnicek (2019) in a similar manner to Lombardozzi and Pitts (2019) draw our attention to the crisis of social reproduction, albeit from a stance rooted in a post-work framework. They point out that many analyses of post-work have been in danger of side stepping questions of social reproduction and productive labour, or to put it differently, the 'really useful labour' that would be required in such a context. Hester and Srnicek (2019) are fearful that this neglect could serve to exacerbate and may even naturalise gender inequalities concerned with care and the nurturing of children. Within a post-work society, despite the rhetoric attached to artificial intelligence, robotisation and the like, there will be a range of tasks that will continue to need to be performed by people. These tasks could range from elder care through to dealing with human detritus and so on. As discussed earlier the thrust of post-capitalist analysis of the forces of production suggests that its development will result in a society of abundance. This is a top down process whereby the development of the forces of production herald the end of capitalism leading to its demise and is reminiscent of Marx and Engels's (1967: 94) description of capital being its own gravedigger. This is a process that follows a logic that is independent of human agency. However, the forces of production are wedded to capitalist relations as is the technology that is mobilised. The point here is that to presume that the development of the forces of production of themselves can herald a post-capitalist society is overly optimistic. This lacuna can be addressed if we reintroduce an analysis of class struggle into conceptualisations of the forces of production. Such an analysis would recognise that there is a class struggle over the application of technology to production which is marked by an antagonistic relationship between labour and capital. This antagonistic relationship is mediated by regional, national and localised contexts as well as by intersectional relations, that is to say, the way in which class, race, gender and so on are interrelated.

For workerist and post-capitalist analyses Marx's discussion of the general intellect in the 'fragments on machines' is of pivotal importance. To reprise the earlier discussion the 'fragments' have been used to analyse the current conjuncture and sits alongside discussions of cognitive or information capitalism etc. Gorz touches on these arguments when he suggests the current conjuncture is qualitatively different from that of industrial capitalism in which the

> main productive force, knowledge, is a product that is in large part, the outcome of an unpaid collective activity, of a 'self production' or 'production of subjectivity'. It is to a large extent, 'general intelligence', shared culture, living and lived practical knowledge'. (Gorz 2010: 52)

In a not dissimilar manner Vercellone (2009) suggests that within cognitive capitalism we need to explore the production of surplus value that arises externally to the capitalist enterprise.

> [Following the gratuitous appropriation of the surplus generated by the social cooperation of labour] it becomes important to rethink the concepts of wage, productive labour and exploitation in a framework where this cooperation is no longer confined within the factory but extended to the whole of society, as it organises itself more autonomously from capital. (Vercellone 2008, unnumbered)

The significance of the above is that it prioritises the development of knowledge, and views this as a collective and implicitly democratic accomplishment occurring outside the direct control of capital. Such arguments are tied to an anthropised analysis of the general intellect and for writers, such as Dyer-Witheford and Pitts, this anthropisation represents a misreading of Marx. Pitts and Dinerstein (2017b: 2) comment,

> Capitalism's specificity pertains not to work but to the forms taken by its results: abstract labour, value, money. Combined, these approaches suggest that the escape from 'work' is no escape route from capitalism.

For Dyer-Witheford et al. the general intellect is represented by its embodiment in capital, whilst for Pitts a better understanding of the current conjuncture can be provided by form analysis allied with social reproduction theory. The point is that these writers refuse a straightforwardly optimistic view of the general intellect and the thrust of capitalist development towards a post-capitalist future. These writers accept that such developments may lead to the absence of waged labour but this does not necessarily represent the transcendence of capitalism. Rather Blacker's description of worklessness may be more appropriate for these writers and to this particular context.

> The current neoliberal mutation of capitalism has evolved beyond the days when the wholesale exploitation of labor under-wrote the world system's expansion. While "normal" business profits plummet and theft-by-finance-rises, capitalism now shifts into a mode of elimination that targets most of us—along with our environment—as waste products awaiting managed disposal. (Blacker 2013: 1)

Dyer-Witheford et al.'s (2019) rather bleak discussion of AI and its potential development echoes Blacker's analysis. Although their analysis is speculative focusing on artificial general intelligence (AGI) they suggest that currently AI cannot match human capabilities. Currently, variable labour power is seen as a particular feature of human labour which is characterised by general intelligence, that is to say, 'the ability to reason with general knowledge, to perform many different tasks, and to operate in new and completely different domains and environments …' (Dyer-Witheford et al. 2019: 126). Nevertheless, Dyer-Witheford et al. speculate that in time AGI will become a possibility thereby rendering human labour redundant and no longer required by capital. The point is that these more pessimistic accounts point to the importance of struggle in shaping societal development and in this way avoid the over optimism surrounding some readings of post-capitalism and accelerationism. They suggest a variety of sites of struggle that could be engaged in the pursuit of a post-capitalist social formation which I discuss below.

Sites of Struggle

Chatterton and Pickerill (2010) in their discussion of anti-capitalist activists not only discuss the range of practices deployed but also point to the fluidity of the strategies mobilised to access a wider constituency. A similar argument could be made in relation to the strategies that could be used in the struggle towards a post-work/post-capitalist society. There are any number of starting points for such a politics, all of which carry with them limitations and omissions. Despite the critique of accelerationism, it does provide a starting point that builds on the development of the forces of production and pushes these to their limit. The omission is that this stance plays down the manner in which technology is enmeshed within capitalist relations. Consequently, it is involved in a twofold struggle. First is to re-appropriate technology for the collective good thereby undermining the way it is shaped to serve capitalist interests and in this way technological advances could be bent to serve all members of society. Second, in its current configuration technology has facilitated processes that have led to deskilling, casualization and precarious labour. This becomes another site of struggle concerned with the availability of decent work. Those capitalist processes that lead to de-skilling, casualization and precarious labour could be both challenged and re-engineered to serve collective ends. Currently, there is a 'schizophrenic' orientation to waged labour, for some, work consumes all of life whereas for others its absence compounds the disadvantages they face. Shortening the working week and increasing the amount of free time available to workers could address this issue and presage alternative conceptualisations of waged labour. If the logic of capitalism is to render increasing numbers of people surplus to its requirements this could presage the possibility of a post-work future, opening-up a politics that moves beyond the preoccupation with waged labour (Harvey 2014). The use of UBI in part addresses this issue, yet may lead to the capitalist state taking on the mantle of the employer with all the negativity that surrounds that. There is a struggle here that needs to move beyond the individualisation that surrounds UBI which needs to connect with collective strategies for social change. Such strategies would engage in discussions of the alternatives to waged labour in post-work/post-capitalist society which need to address the transformation of the social relations of production.

I would like to close this section by referring to Chatterton and Pickerill's (2010) activists whose practices may serve to validate and give social recognition to forms of 'productive' labour that are external to the wage nexus. This of course is another site of struggle that could lead to the development of collective strategies for change that would at the same time address the manner in which socially necessary labour is distributed without compounding existing inequalities. These examples are drawn from what could be described as localised practices. However, such practices need to engage with broader social movements that are concerned with struggles towards change, not only at a local level but also regionally, nationally and globally. The point here is to join with broader struggles that seek to transform and develop socially just societies. Lombardozzi and Pitts (2019) remind us that discussions of waged labour, work and UBI, by failing to move beyond the capitalist social forms in which these are located, are unable to develop a politics that addresses the antagonistic social relations that generate the crisis of social reproduction. These antagonisms derive from the totality of social relations that lie beyond the immediacy of waged and 'free' labour. An adequate politics needs to address the totality of these social relations, failure to do so means critique is lodged within the sphere of social forms that are entwined with capitalist relations. Pitts (2018) argues Postoperaist, or what I have referred to as Workerist, accounts of contemporary work can 'easily dovetail with mainstream accounts of changing capitalism'. We need only reflect on the work and analyses of the WEF and similar institutions.

Vocational Education and Training

In many of the societies of the global north and south, VET is deemed to be of lesser value than that provided by universities (see Cedefop. 2017: 4; Wang and Guo 2019: 551, 553; but see Cedefop. 2017: 24). In the recent past and especially in the UK, VET has been described as being for 'other peoples' children' (Bathmaker 2014a, b). Even when VET is offered in higher education institutions it is frequently devalued, being compared unfavourably with medicine, engineering and so on (Bathmaker 2014a, b). There is a degree of ambiguity in this type of argument that serves to devalue VET, though they do indicate the presence of classed based processes. At the same time universities are increasingly being called upon to be business facing and to prepare their graduates for the world of paid employment. The irony is that elite higher education institutions have historically been pivotal in the reproduction of social relation in terms of class as well as ethnicity. Importantly, attitudes towards VET are mediated by the socio-economic, temporal, spatial and geographic context in which they are placed. In a review Gessler and Siemer (2020: 92) suggest,

> Since the financial crisis of 2007/2008, at the latest, vocational education and training (VET) has gained a high degree of attention, as it is often perceived as a solution to socioeconomic problems following financial crises, particularly the rise of youth unemployment. VET is also viewed as an instrument for increasing an economy's productivity, capacity for innovation and competitiveness.

There are a number of issues that flow from this discussion particularly in an analysis that argues for an expansive construction of VET. Importantly, Esmond points towards the further development of VET and workplace learning in the formation of what he describes as 'technical elites' (Atkins et al. 2019). There is an echo here with discussions of the 4th IR which suggests there is a hollowing out of the occupational structure. Those workers who are described as 'technical elites' able to understand and problem solve creatively when things go wrong will need a holistic understanding of productive processes. In this respect, these elite workers will be required to combine and draw on both 'disciplinary' and practice-based knowledge that underpins production. Access to disciplinary knowledge in this context is not only a feature of distributive justice but also an embedded feature of their labour. It would be risible to ignore the relationship of VET to waged labour and the manner in which for all worker disciplinary knowledge relates to practice and should be thought of dialectically rather than as a static relation. Such a conceptualisation can be related to the significance of occupational and professional responsibilities to wider society, that is their civic responsibilities and social obligations.

Alongside the discussion that addresses the significance of VET for occupational and professional practice there is another debate that has often been marginalised. This debate considers those who lie outside the waged relation and formal economy—the unwaged and workless. It articulates with the social and civic obligations that we owe to one another and is concerned with the development of use values that can contribute towards the well-being of ourselves, the communities within which we live, as well as beyond. It is easy enough to relate this type of VET, particular at this moment of time, with caring forms of labour. However, it extends further than this to encompass a range of self-defined concerns. In the global south these could encompass responses to, as well as the cultural, social and material resources that could be drawn upon to survive poverty and build a 'good' life. In addition VET will be mediated by temporality and our location within our lifespace and lifespan. This would be a rights-based VET that would break free of its erstwhile association with capitalism and a narrow instrumental interest in addressing the needs of industry. In this instance, VET would conjoin with a rights-based education concerned with the furtherance of a socially just society.

Conclusion: The Fourth Industrial Revolution

Conceptualisations of the 4th IR are intertwined with capitalist relations and constitute a deeply ideological notion which carries with it a number of implications for the future of work, social relations, education in general and VET in particular. The term itself could be re-appropriated by a post-capitalist and post-work project. However, such a politics would face a number of obstacles, but no more than any other radical project in the current conjuncture in which the forces of popularism and post-neoliberalism hold sway. Such an engagement would interrogate the 4th IR for the affordances it offers in the development of emancipatory practices. As against those who advocate the distinctiveness of 4th IR who position it as qualitatively different to those of the past, it is important to note that it is nevertheless necessary to hold onto the manner in which industrial revolutions (from the first to the fourth) articulate with capitalism and its accumulation strategies. This engagement, by centring capital, would qualify claims to the 4th IR's distinctiveness. Such a stance would acknowledge that to re-appropriate the 4th IR for an emancipatory project would necessitate a break with its logic which is rooted in capitalist relations.

This chapter addresses arguments that engage with Italian Workerism, post-capitalism, post-work, UBI and paradoxically, the politics of abundance. It seeks to develop a politically informed analysis that moves beyond current conceptualisations of the 4th IR that in turn inform a critical, if not transformative praxis. It serves to remind us that technology is never innocent and that in its current form is conjoined with capitalist interests. Consequently, social and political processes are of central importance in the struggle for change. The affordances provided by the technology surrounding the 4th IR need to be re-appropriated to serve collective interests. This is no easy task, the 4th IR is an ideological construct that can be bent in a number of different directions and can be given a progressive veneer by the apologist of capital. It is for this reason that class struggle is of central importance, but this struggle must be aligned with an expansive understanding of class. Such an understanding would include sections of the middle class as well as those who are involved not only in waged labour but in other forms of 'really useful productive' labour. Care work would be a case in point as would that which contributes towards social reproduction—in this instance the domestic labour of women is pivotal. An expansive understanding of class would also incorporate the workless, those surplus to the requirements of capital who have been excluded from the wage nexus and would similarly necessitate a broader conceptualisation of VET. The struggle for change needs to be informed by a political strategy that operates on a number of fronts, from the local through to the national and global. Whilst such a politics would be located within a revolutionary reformism, it needs to refuse the limitations of such a stance and the danger of its co-optation by a progressive capitalism that seeks to resolve and move beyond the limitation of a post-neo-liberalism.

Acknowledgement This chapter draws in part on my 'Socio-technical imaginary of the fourth industrial revolution and its implications for vocational education and training: a literature review', published in the Journal of Vocational Education & Training, 2018. 70:3, 337–63, https://doi.org/10.1080/13636820.2018.1498907

References

Adler, P. 2007. The Future of Critical Management Studies: A Paleo-Marxist Critique of Labour Process theory. *Organization Studies* 28 (9): 1313–1345. https://doi.org/10.1177/0170840607080743.

Allen, R. 2009. Engels' Pause: Technical Change, Capital Accumulation, and Inequality in the British Industrial Revolution. *Explorations in Economic History* 46: 418–435.

Anderson, D. 2009. Productivism and Ecologism: Changing Dis/courses in TVET. In *Work, Learning and Sustainable Development*, ed. J. Fien, R. Maclean, and Man-Gon Park, 35–57. Dordrecht: Springer.

Anderson, E. (ed.). 2017. *Private Government*. Princeton: Princeton University Press.

Atkins, L. B. Esmond, and R. Suart. 2019. *The Role of Education and Training in the Development of Technical Elites: Work Experience and Vulnerability*. Proceedings of the European Conference on Educational Research (ECER), Vocational Education and Training Network (VETNET). University of Hamburg, Hamburg, September. Berlin: VETNET, 461–468. https://doi.org/10.5281/zenodo.345749.

Autor, D. 2015. Why Are There Still so Many Jobs? the History and Future of Workplace Automation. *The Journal of Economic Perspectives* 29 (3): 3–30. https://doi.org/10.1257/jep.29.3.3.

Avis, J. 2019. A Note on Class, Dispositions and Radical Politics, *Journal for Critical Education Policy Studies* 16 (3): 166–184. http://www.jceps.com/wp-content/uploads/2019/01/16-3-6.pdf.

Bathmaker, A.-M. 2014a, March. *'Applied', 'Technical' and 'Vocational' Constructions of Knowledge in Vocational Education*. Paper presented at Vocational Education and Training: Policy, Pedagogy and Research, BERA Post-Compulsory and Lifelong Learning SIG, The University of Birmingham.

———. 2014b, June. *The Perfect Education for a Person Like Me: Vocational Education and the 'Ordinary', the 'Overlooked' and the Under-Served 40%.* Inaugural Lecture at the University of Birmingham

Beckett, A. 2018. Post-Work: The Radical Idea of a World Without Jobs. *The Guardian.* https://www.theguardian.com/news/2018/jan/19/post-work-the-radical-idea-of-a-world-without-jobs.

Blacker, D. 2013. *The Falling Rate of Learning and the Neoliberal Endgame.* London: Zero books.

Bloodworth, J. 2018. *Hired: Six Months Undercover in Low-Wage Britain.* London: Atlantic books.

Brown, P., H. Lauder, and D. Ashton. 2011. *The Global Auction.* Oxford: Oxford University Press.

Cedefop. 2017. *The Changing Nature and Role of Vocational Education and Training in Europe. Volume 2: Results of a Survey Among European VET Experts.* Luxembourg: Publications Office. Cedefop Research Paper; No 64. https://doi.org/10.2801/548024.

Cederström, C., and P. Fleming. 2012. *Dead Man Working.* Alresford: Zero Books.

Chatterton, P., and J. Pickerill. 2010. Everyday Activism and Transitions Towards Post-Capitalist Worlds. *Transactions of the Institute of British Geographers, NS* 35: 475–490.

Cleaver, H. 2017. *Rupturing the Dialectic.* Chicago: AK Press.

Corbyn, J. 2019. *The Future is Ours to Make, Together.* https://labourlist.org/2019/10/the-future-is-ours-to-make-together-corbyns-campaign-launch-speech/.

Davis, K., and W.E. Moore. 1945. Some Principles of Stratification. *American Sociological Review* 10 (2): 242–249.

Dyer-Witheford, N., A.M. Kjøsen, and J. Steinhoff. 2019. *Inhuman Power.* London: Pluto.

Ehrenreich, B. 2010. *Nickel and Dimed: Undercover in Low-Wage USA.* London: Granta Books.

Fleming, P. 2017. *The Death of Homo Economicus.* London: Pluto.

Frayne, D. 2015. *The Refusal of Work.* London: Zed Books.

———., ed. 2019. *The Work Cure.* Monmouth: PCCS.

Frey, C.B. 2019. *The Technology Trap: Capital, Labor, and Power in the Age of Automation.* Princeton: Princeton University Press.

Fumagalli, A. 2011. Twenty Theses on Contemporary Capitalism (Cognitive Biocapitalism), Angelaki. *Journal of the Theoretical Humanities* 16 (3): 7–17. http://dx.doi.org/10.1080/0969725X.2011.626555.

Gessler, M., and C. Siemer. 2020. Umbrella Review: Methodological Review of Reviews Published in Peer-Reviewed Journals with a Substantial Focus on Vocational Education and Training Research. *International Journal for Research in Vocational Education and Training* 7 (1): 91–125. https://doi.org/10.13152/IJRVET.7.1.5.

Gorz, A. 2010. *The Immaterial*. London: Seagull.

Goulden, C. 2018. *Universal Basic Income—Not the Answer to Poverty*. https://www.jrf.org.uk/blog/universal-basic-income-not-answer-poverty?gclid=EAIaIQobChMIlaCFkd7D5gIVV-DtCh37cwq4EAAYASAAEgKKxfD_BwE.

Graeber, D. 2018. *Bullshit Jobs*. London: Allen Lane.

Haagh, L., and B. Rohregger 2019. *Universal Basic Income Policies and Their Potential for Addressing Health Inequities: Transformative Approaches to a Healthy, Prosperous Life for All*.

Harvey, D. 2014. *Seventeen Contradictions and the End of Capitalism*. London: Profile Books.

Hester, H., and N. Srnicek 2019. *After Work: What Is Left*. https://www.youtube.com/watch?v=fSHT-HKkk8Q.

Hetschko, C., A. Knabe, and R. Schöb. 2014. Changing Identity: Retiring from Unemployment. *The Economic Journal* 124: 149–166. https://doi.org/10.1111/ecoj.12046.

Holloway, J. 2019. *We are the Crisis of Capitalism Oaklands*. Kairos PM.

Hoynes, H., & Rothstein, J. (2019). Universal Basic Income in the US and Advanced Countries NBER Working Paper No. 25538.

Hunnicutt, B. 2013. *Free Time: The Forgotten American Dream*. Philadelphia: Temple University Press.

Kenworth, L. 2004. *Egalitarian Capitalism*. New York: Russell Sage Foundation.

Lombardozzi, L., and F.H. Pitts. 2019. *Social Form, Social Reproduction and Social Policy: Basic Income, Basic Services, Basic Infrastructure*. Capital and Class in press. http://oro.open.ac.uk/62409/1/Lombardozzi%20%26%20Pitts%202019%20Social%20Form%20Social%20Reproduction%20Social%20Policy.pdf.

Mackay, R., and A. Avanessian. 2014. Introduction. In *Accelerate*, ed. R. Mackay and A. Avanessian, 1–50. Falmouth: Urbanomic.

Malloch, M., L. Cairns, K. Evans, and B. O'Connor, eds. 2021. *The SAGE Handbook of Learning and Work*. London: Sage.

Marazzi, C. 2011. *The Violence of Financial Capitalism*. Los Angeles: Semiotext(e).

Martinelli, L. 2017. *Assessing the Case for a Universal Basic Income in the UK*. IPR Policy Brief, Institute for policy research, University of Bath. https://www.bath.ac.uk/publications/assessing-the-case-for-a-universal-basic-income-in-the-uk/.

Marx, K. 1973. *Grundrisse*. Harmondsworth: Penguin.

———. 2014. Fragment on Machines. In *Accelerate*, ed. R. Mackay and A. Avanessian, 52–66. Falmouth: Urbanomic.

Marx, K., and F. Engels 1967 [1872]. *The Communist Manifesto*. Penguin: Harmondsworth.

Mason, P. 2015. *Postcapitalism*. London: Allen Lane.

McGrath, S., and L. Powell. 2016. Skills for Sustainable Development: Transforming Vocational Education and Training Beyond 2015. *International Journal of Educational Development* 50: 12–19.

Painter, A., and C. Thoung. 2015. *Creative Citizen, Creative State: The Principled and Pragmatic Case for a Universal Basic Income*. RSA. https://www.thersa.org/discover/publications-and-articles/reports/basic-income.

Parker, C., and B. Vonow 2017. *Amazon Warehouse Life 'Revealed with Timed Toilet Breaks and Workers Sleeping on Their Feet'*. The Sun. https://www.the-sun.co.uk/news/5004230/amazon-warehouse-working-conditions/.

Pitts, F. 2018. *Critiquing Capitalism Today: New Ways to Read Marx*. London: Palgrave Macmillan.

Pitts, F., and A. Dinerstein. 2017a. Corbynism's Conveyor Belt of Ideas: Postcapitalism and the Politics of Social Reproduction. *Capital & Class* 41 (3): 423–434.

——— 2017b. *Postcapitalism, Basic Income and the End of Work: A Critique and Alternative, Bath Papers in International Development and Wellbeing*. No. 55—November, University of Bath, Centre for Development Studies.

Ponomarenko, V., A. Leist, and L. Chauvel. 2019. Increases in Wellbeing in the Transition to Retirement for the Unemployed: Catching Up with Formerly Employed Persons. *Ageing and Society* 39 (2): 254–276. https://doi.org/10.1017/S0144686X17000976.

Powell, L. 2012. Reimagining the Purpose of VET—Expanding the Capability to Aspire in South African Further Education and Training Students. *International Journal of Educational Development* 32 (5): 643–653.

Rajan, R. 2019. *The Third Pillar*. London: William Collins.

Reich, R. 1992. *Work of Nations*. London: Vintage books.

Rifkin, J. 2014. *The Zero Marginal Cost Society*. London: Palgrave MacMillan.

Rustin, M. 2013. A Relational Society. In *After Neoliberalism? The Kilburn Manifesto*, ed. S. Hall, D. Massey, and M. Rustin. http://lwbooks.co.uk/journals/soundings/pdfs/Soundings%20Manifesto_Rustin.pdf. Accessed 7 Aug 2013.

Sage, D. 2019. The Power of the Work Ethic: Implications for Social Policy. *Critical Social Policy* 39 (2): 205–228.

Shildrick, T., R. MacDonald, C. Webster, and K. Garthwaite. 2012. *Poverty and Insecurity: Life in Low-Pay, No-Pay Britain*. Bristol: Policy Press.

Simmons, R., and R. Thompson. 2011. *NEET Young People and Training for Work: Learning on the Margins*. Stoke on Trent: Trentham Books.

Skidelsky, R. 2016. A Basic Income Could Be the Best Way to Tackle Inequality. *The Guardian*. https://www.theguardian.com/business/2016/jun/23/universal-basic-income-could-be-the-best-way-to-tackle-inequality.

Srnicek, N., and A. Williams. 2015. *Inventing the Future: Postcapitalism and a World without Work*. London: Verso.

Stiglitz, J. 2019. *People, Power and Profits*. Princeton: Princeton University Press.

The Guardian. 2015. *How Sports Direct Effectively Pays Below Minimum Wage*. https://www.theguardian.com/business/2015/dec/09/how-sports-direct-effectively-pays-below-minimum-wage-pay.

Vercellone, C. 2008. The New Articulation of Wages, Rent and Profit in Cognitive Capitalism. In *The Art of Rent*, ed. Queen Mary University School of Business and Management, no page numbers. London. http://hal.archives-ouvertes.fr/docs/00/26/55/84/PDF/The_new_articulation_of_wagesHall1.pdf. Accessed 29 Feb 2012.

———. 2009. Cognitive Capitalism and Models for the Regulation of Wage Relations, The Edu-Factory Collective. In *Towards a Global Autonomous University*, 119–124. New York: Autonomedia.

Wang, A., and D. Guo. 2019. Technical and Vocational Education in China: Enrolment and Socioeconomic Status. *Journal of Vocational Education & Training* 71 (4): 538–555. https://doi.org/10.1080/13636820.2018.1535519.

War on Want. 2019a. *Sweatshops in China*. https://waronwant.org/sweatshops-china.

———. 2019b. *Sweatshops in Bangladesh*. https://waronwant.org/sweatshops-bangladesh.

Weeks, K. 2011. *The Problem with Work*. Durham: Duke University press.

Wilkinson, R., and K. Pickett. 2010. *The Spirit Level: Why Equality Is Better for Everyone*. Harmondsworth: Penguin.

———. 2018. *The Inner Level, How More Equal Societies Reduce Stress, Restore Sanity and Improve Everyone's Well-Being*. Milton Keynes: Allen Lane.

World Economic Forum. 2018. *The Future of Jobs 2018*. Geneva: WEF. http://www3.weforum.org/docs/WEF_Future_of_Jobs_2018.pdf.

World Health Organization. 2019. Universal basic income policies and their potential for addressing health inequities: Transformative approaches to a healthy, prosperous life for all. http://www.euro.who.int/en/health-topics/health-determinants/social-determinants/publications/2019/universal-basic-income-policies-and-their-potential-for-addressing-health-inequities-transformative-approaches-to-a-healthy,-prosperous-life-for-all-2019.

Zukas, M., and J. Malcolm. 2015. *Learning to Be a Social Scientist*. Paper Presented RWL9—9th International Conference on Researching Work and Learning, Singapore, December 9–11.

5

Conclusion: Vocational Education in the Fourth Industrial Revolution—Education and Employment in a Post-Work Age

Abstract This chapter brings together the key themes of the book drawing out their significance for the development of vocational education and training and education systems committed to a socially just society. Such interventions need to go beyond the narrowly educative and engage with a broader politics committed to the transformation of society. Many of the rhetorical claims surrounding the fourth Industrial Revolution (4th IR) align themselves with a notion of social justice. However, this is set within a capitalist terrain that seeks to manage the tensions flowing from the conditions presaged by the 4th IR, rather than their transcendence. Such rhetorical claims frequently serve to domesticate tensions thereby underplaying their potential in the struggle for the creation of a socially just society.

Keywords Fourth Industrial Revolution • Vocational Education and Training • Social Contract • Neo-liberalism • Transformism • Social justice

J. Avis, *Vocational Education in the Fourth Industrial Revolution*,
https://doi.org/10.1007/978-3-030-52032-8_5

Introduction

This chapter serves as a conclusion, engaging with a number of themes discussed earlier and that draws out the significance of the fourth Industrial Revolution (4th IR) for vocational education and training (VET) and education systems more generally. A key concern is the relationship of educational processes and their articulation with the aspiration for a socially just society. If such an aspiration is taken seriously it would demand interventions that go beyond the narrowly educative and would need to engage with a broader politics that seeks the transformation of society. Such a politics would necessarily be anti-capitalist as existing capitalist relations fetter the development of socially just societies. Paradoxically the crisis and failings of neo-liberalism are widely discussed and recognised. The overarching theme of the 2020 Davos conference is 'stakeholders for a cohesive and sustainable world'. Schwab (2020: online) in his introductory comments to the conference referred to the failings of the economic elite and, by default, neo-liberalism.

> People are revolting against the economic elites they believe have betrayed them, and our efforts to keep global warming to 1.5 °C are falling dangerously short, with the world at such critical crossroads, this year [at the conference] we must develop a 'Davos Manifesto 2020' to reimagine the purpose and scorecards for companies and governments.

He anticipates that the development of some sort of scorecard (reminiscent of those used by the European Commission 2019) would be able to call companies and governments to account and thereby lead to change. Concerns about the abuses of economic elites, the lack of fairness throughout the global south and north, together with allied worries about global warming and the climate emergency are ubiquitous. The critique developed by those who could be described as the apologists of capital anticipate the development of a new social and economic settlement reminiscent of stakeholder capitalism (Avis 1998), one that seeks to restore stability to the system. The inequalities perpetuated by neo-liberalism and the potential tensions and conflicts engendered pose a threat to the system and its legitimacy. Notably, the celebration of the market and competition allied with the privatisation of parts of the welfare state and utilities, rather than benefitting society as a whole, are seen to be fraudulent. In addition, wages have been decoupled from productivity and growth leading to an incoherent distribution of both income and wealth. Neo-liberal rhetoric has been shown to be just that, mere rhetoric. In effect, the conjuncture is marked by a post-neo-liberalism which carries with it a deep-set cynicism, whereby the neo-liberal subject, is no longer held in check by social constraints. This applies particularly to the excesses of financialisation and rentier capitalism and can be seen in the bonuses paid to senior executives whose companies fail. The call for a new social contract, or what I am referring to as a settlement, aims to respond to these new conditions, seeking to restore legitimacy to the system. The development of a new social contract in part reflects societal instability and in part a shift in power away from capital towards labour. Implicitly these concerns are indicative of a fear of disorder and emerging class struggle that 'progressive' capitalism encounters. This leads to a potential class grouping that struggles against the irrationality, unfairness and irresponsibility of a post-neo-liberalism but that simultaneously seeks to leave in place hierarchical and capitalist relations.

The conceptualisation of the 4th IR constitutes a socio-technical imaginary. It is a rhetorical notion that nevertheless has material and social consequences and unsurprisingly is rooted within capitalist relations. We could trace a relationship between it, the demise of Fordism, the shift to post-Fordism and the ascendancy of neo-liberalism in the 1970s. These shifts whilst uneven can be related to the balance of power between capital and labour and therefore to the outcome of class struggle. Much is made of the dislocations that flow from the 4th IR, many of which articulate with the failings of a post-neo-liberalism present in the current conjuncture.

The dislocations of the 4th IR incorporate a range of tensions and contradictions which include the polarisation not only of income and wealth but also of jobs between the well and badly paid, the good and lousy, and so on. This is allied to processes of de- and up-skilling which has resulted in increased precariousness and a surplus population excluded from waged labour. These dislocations have impacted upon waged labour in several ways. For many increased precariousness and the fear of unemployment have resulted in the intensification of labour, the extension of the working day whereby all of life is put to work. This process is particularly acute for members of the professional middle class who find themselves employing domestic workers to care for their children, elderly parents and to clean their homes whilst they are at work. Care work is notoriously badly paid and such employees may find themselves taking on multiple jobs so as to provide for their families. In addition, there are those who live in disadvantaged regions with few opportunities for waged employment and who consequently churn between low-waged and intermittent employment, interspersed with periods of unemployment. Such processes represent a crisis in social reproduction. The concern to develop a new social contract or settlement by particular fractions of capital represents an attempt to manage and overcome this crisis, thereby restoring stability to the system.

The spectre of technological unemployment haunts discussions of the labour market and 4th IR. Whilst labour economists refuse this claim suggesting it is mere speculation out of kilter with the historical record, the logic embedded within the 4th IR and indeed capitalist development, is towards the expulsion of workers from waged employment. However, this logic is softened through the use of 'living' labour power to augment AI. This argument is supported by the claim that the efficiencies of both 'living' labour power and AI are enhanced when they work together. Chess is often used as an example to argue that a blend of AI and human labour power is more successful than AI or labour power on its own. This derives from labour's capacity for general intelligence, the ability to reason and to operate in different environments. The continued significance of labour in augmenting AI would remain the case, until and if ever, artificial general intelligence is able to overcome the obstacles it faces and be fully developed. Yet at the same time the use of AI to augment labour power may also be a gloss for deskilling. It is also important to acknowledge that the costs incurred to create 'smart' factories may be such as to render less efficient and more labour intensive manufacturing processes to be economically viable. This would lead to what Fleming (2017: 163) describes as over-employment by which he has in mind workers performing processes that can be readily automated. Workers could be freed from such tasks but in the above example are not replaced for economic reasons.

The intensification of labour and its precariousness have resulted in a crisis of social reproduction. This crisis is reflected in the lived experiences of those who find it necessary to sell their labour power and includes those dependent on a wage, ranging from the professional middle class to the care workers they employ. Marx (1976: Chapter 10) draws our attention to the damaging consequences of the extension of the working day—the intensification of labour for all involved from the worker who encounters the danger of accidental injury to the factory owner who consequently faces declining profits as a result of interruptions to production. The restriction of the working day was a response to the then crisis of social reproduction. Paradoxically, the same issues arise today. As in Marx's day the current crisis of social reproduction also impacts on surplus labour—those who have been excluded from waged employment and are no longer required by capital.

It is important to acknowledge uneven development, that is to say, if we were to consider the 4th IR as a mode of production it exists alongside earlier modes which may operate with a slightly different capitalist logic. The 4th IR and its relationship with AI, digitalisation, robotisation and so on, carries with it an exclusionary logic whereby labour is expelled from the workforce and sits alongside processes of de-industrialisation, the decline of Fordism in the US, the UK and elsewhere in the 1970s and 1980s. This led to the growth of disadvantaged communities that had lost their 'industrial' base and with this, employment opportunities. The long-term interest in VET, technical education and retraining is a frequently repeated policy response to those earlier conditions as well as to the current conjuncture. The aim being that such interventions would serve to open up the labour market for those who were formally excluded. Whilst such interventions seek to address the crisis of social reproduction, through their individualisation and concern to prepare participants for intermittent low-skilled and waged employment do little to interrupt disadvantage or indeed address this crisis. In this policy response there are echoes of social democracy, egalitarian and also a 'progressive' capitalism. Within the rhetoric of 'progressive' capitalism the interest is not only in providing employment possibilities but also to address, or rather ameliorate, poverty and low wages without challenging the structural relations within which disadvantage is set.

There is nothing particularly radical about the preceding paragraphs, but they can lead to a discussion of the extension of free time as well as to debates surrounding universal basic income (UBI), both of which can be used to challenge low-waged employment. AI can be used to automate formerly low-skilled and low-waged labour and could thus lead to a reduction in the hours of work and an expansion of free time. In the case of the latter, by providing a basic income workers will no longer be forced to undertake low-waged work. In addition, a guaranteed income will ameliorate the conditions that many workers face, characterised by precariousness, intermittent employment, the insecurities surrounding the 'gig' economy and a fluctuating income that makes it difficult to manage finances.

Whilst UBI may offer a number of progressive affordances, it does not necessarily challenge the social relations of production or capitalism. Indeed, UBI could readily accommodate a progressive capitalism that seeks to address some of the dislocations that have been generated by a rampant neo-liberalism. Such an accommodation could be orchestrated by supranational organisation such as the WEF, World Bank and so on, echoing the practices of the state in managing the contradictions of capitalism for the benefit not of all members of society, but of capital. In this instance, these supranational organisations are not solely acting on a local or regional scale, but also globally. The recognition of diversity, equity, human-centred growth and long-term sustainability, together with the development of circular economies that seek to avoid waste through the continual re-use of resources, whilst located on a capitalist terrain nevertheless have progressive possibilities. These concerns could be conjoined with a revolutionary reformism but are also amenable to appropriation by capital. For example, the interest in human-centred growth, the development of circular economies and long-term sustainability, all address the climate emergency but can be readily co-opted by capital in its endless pursuit of new markets. I am reminded of processes of 'creative destruction' (Schumpeter 1994 [1943] chapter vii) as well as Marx and Engel's (1985: 83) comment that 'all that is solid melts into air'. Environmentally sound manufacturing, the development of solar power allied with the Internet of Thingsand so on could all serve to reinvigorate capitalism. There is a transformist thrust to these arguments whereby 'fundamental' change is set on a terrain that seeks to contain popular dissent and secure the interests of a particular fraction of capital—a passive revolution (Gramsci 1971: 106–114). At the same time the concern to address the climate emergency and the precariousness of waged labour are sites of class struggle. The paradox is that the affordances of the 4th IR and the related technologies have all been shaped by but also held in check by capitalist relations. In this respect, accelerationists are correct in arguing that the development of the forces of production could portend a post-capitalist and post-work social formation but this possibility is constrained by the manner in which such affordances are articulated with capitalist interests. This is a site of struggle in which education and VET can play a part.

The earlier discussion in this chapter has a bearing upon the way in which we could think about the salience of education and in particular (T)VET in the current conditions, particularly those surrounding the uncertainties of and exclusion from waged labour. It is easy enough to argue that in a knowledge-orientated society VET needs to support and develop the expertise and skills required. However, if we take seriously commitments towards the development of a socially just society, VET and education in general needs to address the responsibilities owed to wider society and not solely privilege the needs of industry—which is nearly always a gloss for the needs of capital (Brockmann et al. 2011). We need to ask ourselves what are the skills, knowledge and expertise needed in the current conjuncture. This cannot be determined by fiat but has to be collectively and democratically decided at local, regional, national and global levels with all the difficulties this involves. In the same way as VET has responsibilities to wider society so too do occupations and professions, but in both cases this more correctly should be conceived not solely locally but also globally. In this way the inequalities that traverse the high-, medium- and low-income economies would be brought into view, thereby calling into account the manner in which these are reproduced and the way in which they could be addressed. Whilst supranational organisations call for human-centred growth, the provision of decent work and gender equality we need to consider the way in which equality is understood. Neo-liberalism individualises questions of inequality and works with constructions of the atomised competitive subject. We should break with such notions replacing these with collective understandings which would undermine functionalist notions of inequality that portray the successful as having rare and remarkable skills. A break with such conceptualisations would serve to justify a flatter distribution of income and wealth, which in turn could facilitate the development of a more prosperous and equitable society. This returns us to education and VET and the need to interrogate current practices for their ability to benefit all members of society and enable the development of 'really useful labour'. Much is made of the importance of 'soft' human and cognitive skills by the senior executives canvassed by the WEF (see Chaps. 4, 6 and 7). These skills include the ability to communicate, persuade, problem solve, interpret data and so on, which should be addressed and developed by education and VET. These skills are Janus-like and could be domesticated and bent to capitalist interests yet at the same time mobilised by those at the bottom of the occupational structure in pursuit of a socially just society.

We express our species being through 'productive' labour which may arise within or external to the wage nexus. In the former, this is undermined by capitalist relations and in the latter is frequently overlooked or marginalised. In societies in which surplus labour and the growth of worklessness are becoming more prevalent or are marked by low wages and in-work poverty, the issues of surplus labour and over-employment are increasingly apparent. In some respects, these questions are addressed by post- and anti-work analyses but here I want to consider the relationship of these to VET. If we take an expansive view of VET as being concerned with the social development of labour, this can serve to break free from a narrow association with paid employment and the needs of capital. Somewhat idealistically this can be linked to Marx's notion of unalienated labour as well as to notions of human development. An expansive understanding of VET may provide an engagement with forms of really useful and powerful knowledge that, freed from their association with capitalism, presage the possibility of a socially just society. Such possibilities arise not from the linear development of society or the forces of production, but are the result of struggle and the balance of power between capital and labour. In the latter case, this operates with an expansive understanding of class that acknowledges intersectionality, the articulation of class with race and gender, and includes not only the workless and unwaged but also sections of the middle class.

Throughout this book I have been concerned with an analysis of VET, the manner in which it is conceived as well as the issues it should address. I have avoided a direct engagement with its institutional forms. For example, I have not compared the German dual system with that found in Switzerland or its suitability for the US and so on (Emmenegger et al. 2019; Gessler and Peters 2017). Nor have I considered VETs expression in co-ordinated market economies compared with liberal market economies (Bosch and Charest 2008). There are several reasons for this, not least amongst which would be that this would call for a very different book. In addition VET by its very nature is diverse and dynamic, subject to ongoing change. Whilst the UK is perhaps the best example of inchoate change marked by stops and starts (see Keep 2006), VET systems elsewhere are also subject to continual change. As Cedefop (2017a: 5) points out,

> Vocational education and training takes many forms; it is the most heterogeneous of the main education and training sectors in Europe today. It is difficult to grasp VET as a single institutional entity as it overlaps with other parts of the education and training system in many cases.

This diversity not only exists across Europe but also within national systems which in turn are marked by regional, temporal and sectoral differences (Cedefop 2017a: 10). Thelen and Busemeyer (2008) draw our attention to the shift from collectivism towards segmentalism in the German VET system.

> In collectivism employers were encouraged to over train thereby producing workers with broad and portable occupational skills whereas in the latter [segmentalism] training is organised around internal labour markets and the specific needs of the companies concerned. (Thelen and Busemeyer 2011: 69)

The point is, VET is always set within socio-economic and historical conditions, with the shift from collectivism to segmentalism reflecting a change in the balance of power between labour and capital. In addition, there are many other social-economic changes that have impacted upon VET and its institutional forms. Such changes would include the following: the growth of youth unemployment leading in some cases to the warehousing of young people (Avis et al. 2016, 2017); a rethinking of VET away from seeing it as a 'dead-end initial training for skilled workers, clearly separated from general education' (Cedefop 2017b: 4). In addition, there have been attempts to align VET with lifelong learning as well as the development of hybrid forms. These seek to bring together VET with general education, providing routes to higher education.

Regardless of the specificity of the institutional and systemic contexts within which VET is placed, there is a shared understanding of what it constitutes. Cedefop (2014: 292) defines VET as 'education and training which aims to equip people with knowledge, know-how, skills and/or competences required in particular occupations or more broadly in the labour market' (Cedefop 2014: 292 and see Cedefop 2017a: 5). Such a conceptualisation is reflected in those of the expert reviewers canvassed by Cedefop (2017b: 4), 'Despite the variety, VET is largely perceived by the experts consulted as occupations-specific education and training geared towards securing supply of skilled labour'. In many respects these views are unsurprising, almost by definition VET is orientated towards the labour market and employment, but this represents a truncated and limited view of what VET could presage.

VET is multifaceted and points in any number of directions ranging from its role in warehousing young people to its preparation of technical elites (Avis et al. 2016; Atkins et al. 2019). It has the potential to address the needs of those who are placed outside the wage nexus and who are involved in really useful labour and contribute to the development of what we may describe as use values. This potentiality arises not just in the global north but also in the south applying in particular to those who are nominally workless and living in poverty—the development of use values that contribute to the well-being of communities. In some respects, these practices have been prefigured by community and adult education, but more often than not such practices are underpinned by capitalist relations. The point is to push to the limit such practices with an aspiration towards a socially just society thereby seeking to develop a fully expansive understanding of VET—a conceptualisation that goes beyond one based on a narrow occupational and labour market understanding.

Conclusion

The contradictions surrounding the current conjuncture provide the material basis from which the struggle for a transformation of society could arise, a context in which the move towards post-capitalist and post-work social formations is a possibility. Whilst these affordances can be seen in the technologies and related struggles surrounding the 4th IR, they are constrained and limited by the way in which the 4th IR is embedded within capitalism and its accumulation strategies. AI is pivotal to these processes and together with the related technologies can contribute to both the accumulation of capital and a declining rate of profit. In the former AI could serve as a general condition of production, an infrastructural support that benefits capital in a similar fashion to electricity or transport systems. In the case of the latter the expulsion of labour from waged work could lead to a declining rate of profit with financialisation seeking to bypass production and the declining rate of profit by using money to make money. However, this strategy in turn creates its own contradictions and limits because of its failure to add and produce value.

Whilst revolutionary reformism, as a result of its model of gradual change, is open to co-option, it does nevertheless serve as a starting point for struggle. The limits of post-neo-liberalism as well as the monopolistic and oligarchic tendencies of capitalist organisations give rise to dissent which derive from experiences of precariousness, exclusions as well as the oppressions of waged labour and so on. The attempt to form a new social contract reflects an endeavour by capital to manage these tensions. However, these attempts can be turned back on themselves and contribute towards anti-capitalist struggles. The social contract's concern with the development of human capital as well as the provision of decent and sustainable work can be turned back on itself when these aspirations are compared with the lived experiences of waged labour and neo-liberalism. The recent history of capitalism reminds us of its opportunism which is illustrated in the shift from social democracy to forms of neo-liberalism. When it was auspicious for capital the social democratic concessions won by the working class were withdrawn, being replaced by the ruthlessness of neo-liberalism. Localised anti-work struggles prefigure alternative organisational forms to those of capital, pointing towards progressive possibilities. There are any number of sites of anti-capitalist struggle which can be occupied and mobilised in the pursuit of social transformation. Despite the blandishments that surround the 4th IR, it is firmly embedded in capitalist relations and history reminds us that a reformed human-centred capitalism is nevertheless a tautology.

Acknowledgement This chapter draws in part on my 'Socio-technical imaginary of the fourth industrial revolution and its implications for vocational education and training: a literature review', published in the Journal of Vocational Education & Training, 2018. 70:3, 337–363, https://doi.org/10.1080/13636820.2018.1498907.

References

Atkins, L., B. Esmond, and R. Suart. 2019. *The Role of Education and Training in the Development of Technical Elites: Work Experience and Vulnerability.* Proceedings of the European Conference on Educational Research (ECER), Vocational Education and Training Network (VETNET). University of Hamburg, Hamburg, September. Berlin: VETNET, 461–468. https://doi.org/10.5281/zenodo.345749.

Avis, J. 1998. (Im)possible Dreams: Post-Fordism, Stakeholding and Post-Compulsory Education. *Journal of Education Policy* 13 (2): 251–163.

Avis, J, K. Orr, and P. Warmington. 2016. *Race, 'Warehousing' and Vocational Education and Training: a European Issue?* Paper presented at ECER, University College Dublin, September 23–26.

———. 2017. Race and Vocational Education and Training in England. *Journal of Vocational Education & Training* 69 (3): 292–310. https://doi.org/10.1080/13636820.2017.1289551.

Bosch, G., and J. Charest. 2008. Vocational Training and the Labour Market in Liberal and Co-Ordinated Economies. *Industrial Relations Journal* 39 (5): 428–447.

Brockmann, M., L. Clarke, and C. Winch. 2011. *Knowledge, Skills and Competence in the European Labour Market*. London: Routledge.

Cedefop. 2014. *Terminology of European Education and Training Policy: A Selection of 130 Key Terms*, 2nd ed. Luxembourg: Publications Office. http://www.cedefop.europa.eu/files/4117_en.pdf.

———. 2017a. *The Changing Nature and Role of Vocational Education and Training in Europe. Volume 1: Conceptions of Vocational Education and Training: An Analytical Framework*. Luxembourg: Publications Office. Cedefop research paper; No 63. https://doi.org/10.2801/532605.

———. 2017b. *The Changing Nature and Role of Vocational Education and Training in Europe. Volume 2: Results of a Survey Among European VET Experts*. Luxembourg: Publications Office. Cedefop research paper; No 64. https://doi.org/10.2801/548024.

Emmenegger, P., L. Graf, and C. Trampusch. 2019. The Governance of Decentralised Cooperation in Collective Training Systems: A Review and Conceptualisation. *Journal of Vocational Education & Training* 71 (1): 21–45. https://doi.org/10.1080/13636820.2018.1498906.

European Commission. 2019. *European Innovation Scoreboard 2019*. https://ec.europa.eu/growth/industry/innovation/facts-figures/scoreboards/.

Fleming, P. 2017. *The Death of Homo Economicus*. London: Pluto.

Gessler, M., and S. Peters. 2017. Implementation of Dual Training Programmes through the Development of Boundary Objects: A Case Study. *Educar* 53 (2): 309–331.

Gramsci, A. 1971. *Selections from the Prison Notebooks*. London: Lawrence and Wishart.

Keep, E. 2006. State Control of the English Education and Training System—Playing with the Biggest Train Set in the World. *Journal of Vocational Education & Training*. 58 (1): 47–64. https://doi.org/10.1080/13636820500505819.

Marx, K. 1976 [1867]. *Chapter 10 Capital Volume 1*, 340–416. London: Penguin.

Marx, K., and F. Engels. 1985 [1888]. *The Communist Manifesto*. London: Penguin.

Schumpeter, J. 1994 [1943]. *Capitalism, Socialism and Democracy*. London: Routledge.

Schwab, K. 2020. https://www.weforum.org/events/world-economic-forum-annual-meeting-2020/about.

Thelen, K., and M. Busemeyer. 2008. *From Collectivism Towards Segmentalism: Institutional Change in German Vocational Training*. MPIFG discussion paper 08/13, Coln, Max-Planck-Institute.

———. 2011. Institutional Change in German Vocational Training: From Collectivism Towards Segmentalism. In *The Political Economy of Collective Skill Formation*, ed. M. Busemeyer and C. Trampusch, 68–100. Oxford: Oxford University Press.

Postscript COVID-19

I am writing this postscript whilst engaged in the final review of this book's manuscript. It remains to be seen whether COVID-19 both represents and will lead to a fundamental transformation in social and economic relations. It may, however, accelerate the shift of economic power away from the west towards Asia, as a result of the equivocations of the Trump administration and the EU and may also exacerbate the long-term decline of the UK (Tooze 2020). The secular impact of the pandemic on the labour market, vocational education and training (VET) and wider society is yet to be ascertained.

Earlier in the book I drew on Gramsci's notion of an interregnum whereby 'the old is dying and the new cannot be born' (Gramsci 1971: 276). The erstwhile hegemony of neo-liberalism has been thoroughly debunked—the tenets upon which it was built have been shown to be fallacious. The claims of the superiority of the market, the furtherance of competition across all facets of social life, the privatisation of utilities and parts of the welfare state have all been shown to exacerbate inequality. To the extent that neo-liberalism is hegemonic, it is in the form of a post-neo-liberalism that is characterised by a deep-set cynicism founded upon

J. Avis, *Vocational Education in the Fourth Industrial Revolution*,
https://doi.org/10.1007/978-3-030-52032-8

low trust, the pursuits of elites to secure their own interests, and a rampant capitalism (Dorling 2020: 181) marked by the decoupling of employment and wages from productivity and growth. This disconnect is reflected in a culture in which senior managers feel empowered to pay themselves large bonuses even though their companies are failing or underperforming. Perhaps the best example of the failings of neo-liberalism can be seen in finance capitalism, the mobilisation of algorithms and the use of money to make money. In the early stages of the COVID-19 crisis, the rapid ups and downs of the world's stock exchanges enabled finance capitalist to gain easy profits. Similarly, the world shortages of ventilators and protective equipment during the pandemic have provided scope not only for acts of solidarity and generosity but also for the pursuit of profit. Butler (2020, online) in her blog addressing COVID-19 refers to

> the arrival of entrepreneurs eager to capitalize on global suffering, all testify to the rapidity with which radical inequality, which includes nationalism, white supremacy, violence against women, queer, and trans people, and capitalist exploitation find ways to reproduce and strengthen their powers within the pandemic zones.

The current COVID-19 crisis draws our attention to the inadequacies of all forms of neo-liberalism. Yet it is important not to reify COVID-19 and to attribute to it an undeserved determinism. The way in which the current crisis surrounding the pandemic plays out will be the outcome of struggle. Both Agamben (2020) and Klein discuss the manner in which the current crisis will be mobilised by government and economic elites to secure their own interests. Such practices are a feature of Klein's conceptualisation of the 'shock doctrine' and disaster capitalism. In her VICE interview, she comments (Solis 2020, online; and see Klein 2007), 'The "shock doctrine" is the political strategy of using large-scale crises to push through policies that systematically deepen inequality, enrich elites, and undercut everyone else'. Importantly, she adds, 'In moments of crisis, people tend to focus on the daily emergencies of surviving that crisis, whatever it is, and tend to put too much trust in those in power. We take our eyes off the ball a little bit in moments of crisis'. She alerts us both to vigilance and by implication the contestations surrounding crises. Agamben, in a related argument, suggests that the crisis will intensify and further the development of the surveillance state (and see Zuboff 2019). He comments (2020, online) that

> … security governments do not necessarily work by producing the exceptional situation, but by exploiting and managing it when it occurs. I am certainly not the only one who thinks that for a totalitarian government like China, the epidemic was the ideal way to test the possibility of isolating and controlling an entire region. And the fact that in Europe we can refer to China as a role model shows the degree of political irresponsibility to which fear has brought us.

This tension becomes a site of struggle although some commentators have suggested that Agamben views the COVID-19 crisis as presaging the development of a neo-fascist state (Cohen 2020, online). This arises as a result of the enhancement of the surveillance structure of the state that originated, at least in the west, in the concern to monitor the pandemic. Agamben fears that such a structure will be left in place with all the dangers this augurs.

Currently the credo of austerity and tenets of neo-liberalism have been pushed to one side, being replaced by an interventionist state seeking to secure the long-term interests of capital. In the UK this state mirrors a form of one nation conservativism in which we are all constructed as part of the same community, despite our differences. However, it is important to acknowledge that the parameters of what constitutes community are both imaginary and contested. We could, for example, conceive of a post-capitalist state which would question the inclusiveness of one nation conservativism with its authoritarian underpinnings. Simultaneously, the importance of the welfare state and the nursing and social care workforce for societal well-being is being forcibly brought to our attention in the current crisis—that is to say the work of those who labour as care workers, refuse collectors and of course those working in the National Health Service. Care workers in particular who have previously been undervalued and offered low wages are now recast as 'heroes' engaged in what may be described as 'really useful' labour but also life-threatening work. These carers may be waged but are often unwaged—we need only think of the contribution of voluntary workers whose practices could be seen to prefigure a transformation of social and economic relations towards a more caring society. The current crisis points towards a contrast between what can be described as 'really useful' labour set against the 'really useless' labour of those involved in finance capitalism. There are other tensions and contradictions surrounding current social and economic relations not least the relatively secure incomes of senior managers set against the wage cuts, redundancies and furloughs faced by low-waged manual and white-collar workers. The key tension remains between the logic of capitalism, its pursuit of accumulation and the salience of really useful labour.

Throughout this book I have been concerned with a number of issues:

- the significance of the imaginary of the 4th IR,
- the construction of the economy,
- the salience of digitalisation,
- the logic of capital towards the expulsion of labour from production (Harvey 2014: 104–111).

I have sought to consider the implications of these processes for vocational education and training. It is important to recall that the logic of capital towards the expulsion of labour from production could be and has been moderated by economic calculation—if the cost of replacing labour by machinery was greater than the cost of low-waged workers the latter would remain in place. These processes are set within a context marked by the hollowing out of the labour market and the automation of routine manual and non-manual labour/tasks. At the same time the significance of and need for highly skilled 'elite' technical, professional and managerial workers coexists alongside the displacement of labour.

To the extent that the above notions describe what was the then current context, it is one derived from the balance of power between labour and capital in which the later has been dominant. In this instance an expansive notion of labour and class is mobilised that acknowledges not only the waged but also the unwaged, under- and over-employed as well as those who have been expelled from the formal economy—the workless. In this context, VET is to address the high-level skills required by 'elite' technical, professional and managerial workers as well as the putative development of a flexible adaptive labour force suited to the gig/service economy who encounter periods of under/unemployment.

So how does all of this sit with COVID-19? The current conjuncture draws our attention to the profiteering of particular sections of capital but also alerts us to the importance of the nursing and social care workforce. We should include within this workforce those who are placed outside the wage nexus. How then should VET respond to the needs of this workforce? In addition, COVID-19 serves to remind us of the corporate responsibilities organisations have to wider society as well as our social obligations. Much the same can be said about the responsibilities of occupations and professions to wider society and the manner in which these ethical concerns can be aligned with VET. Importantly, these ethical responsibilities are subject to ongoing dialogue and development. VET needs to move beyond the form of expansive learning that Fuller and Unwin (2003) propose to one that would enable a full-fledged critique of capitalism. However, this critique would not be subject to the assertion of a dogmatic diktat but rather on the basis of dialogue and discussion. It is important, however, to acknowledge that education in general and VET in particular 'plays a key role in the perpetuation of the capital relation; this is the skeleton in capitalist education's dank basement' as Allman, McLaren and Rikowski remind us (2003: 149). The implication is that VET is capitalist education with practitioners and participants often struggling to relate it to wider social relations and seeking to provide a dialogic engagement with 'really useful' knowledge. 'Gig' workers, the intermittently employed who engage in shifting patterns of employment raise a number of thorny question for VET. Perhaps in the current context of COVID-19 in which the state has pushed to one side fiscal constraint, the significance of VET and education more generally as a right for all could be reasserted. Much the same point could be made in relation to those who have been expelled from the workforce and find themselves located in the informal economy—the workless. All of this serves to raise the question of what does it mean to be a vocationally educated member of society which in turn returns us to VET, its responsibilities to wider society as well as our social obligations to one another.

We could envisage a number of scenarios post-COVID-19 which themselves would become sites of struggle, some of which have been alluded to earlier in this postscript. Although Dorling (2020) in his book is not discussing COVID-19, his views were reported in the *Times Higher Education Supplement* (Reisz 2020: 16). He argued that the earlier history of capitalist development was marked by the rapidity and acceleration of transformative change which continued until the 1970s. These changes carried in their wake the despoliation of the natural world, the intensification of inequality and all the abuses that we associate with a crass neoliberalism or, in Dorling's terms, a rampant capitalism. However, the rapidity of change has slowed and decelerated. The pandemic in some senses prefigures a slower world and paradoxically the potential benefits that derive from it—that is not withstanding the inequalities that it has spawned and deepened. Berry (2020, online) suggests that at least some on the left 'hope that from the ashes of this crisis the foundations of a better society will rise'. Žižek (2020) has called for a reinvented communism that recognises our interdependencies and collective responsibilities. This is an argument that resonates with that of the Care Collective (2020, online) who argues that the pandemic is not solely 'the result of a new pathogen circulating around the world. It is also a crisis of care'. The resolution is to place care at the centre of social relations and organise society accordingly. Responses to the pandemic alert us to the importance of care and our social obligations to one another and prefigure an alternative way of ordering society. A more sustainable and slower world is presaged in Dorling's writing but also in the experiences of lockdown in the major cities of the world. Gambuto (2020, online) writes,

> What the trauma has shown us, though, cannot be unseen. A carless Los Angeles has clear blue skies as pollution has simply stopped. In a quiet New York, you can hear the birds chirp in the middle of Madison Avenue. Coyotes have been spotted on the Golden Gate Bridge. These are the postcard images of what the world might be like if we could find a way to have a less deadly daily effect on the planet.

Such experiences prefigure the development of a greener, environmentally sustainable and more equitable politics and could revive or provide added impetus to 'green new deals' (Galvin and Healy 2020; The Green New Deal Group 2008; Powell 2019; Sierra Club n.d.-a, -b). Such a politics would seek to address environmental issues but would also aim to rectify the excesses of inequality that occur in capitalist states. The Sierra Club (n.d.-b) calls for a '*people's bailout*' which highlights the provision of health care, respect for the environment and calls for a version of universal basic income and the redirection of resources away from the rich to the communities and those hardest hit by the pandemic. In the US this would include Native Americans and communities of colour as well as sections of the white working class. Similarly, in the UK it is ethnic minorities who have been hit hardest by the pandemic, reflecting their position in the labour market.

As against the image of a silent and peaceful city Gambuto (2020, online) reminds us of other scenes which are echoed in the UK,

> a health care system that cannot provide basic protective equipment for its frontline; small businesses—and very large ones—that do not have enough cash to pay their rent or workers, sending over 16 million people to seek unemployment benefits; a government that has so severely damaged the credibility of our media that 300 million people don't know who to listen to for basic facts that can save their lives.

Aftermath: The New Normal

Many commentators have argued that the pandemic will fundamentally transform social, political and economic relations. These stances vary between those adopting a pessimistic position such as Agamben, who argue the pandemic augurs the development of a neo-fascist surveillance state. Others have argued that that the pandemic opens up the possibility for the development of a kinder, more egalitarian socialist society (Dorling 2020; Žižek 2020). Whatever the outcome, this will be shaped by the balance of power between labour and capital as well as by the progression of the pandemic. Cohen (2020), drawing on actor network theory, would describe the virus as an actant which has its own agentic properties. It is worthwhile acknowledging this possibility, but putting this to one side and assuming that the pandemic is controllable, it is likely that in the immediate aftermath of the crisis there will be strenuous attempts by capital to return to the state of affairs that preceded it. Gambuto (2020) imagines that vast sums of money will be spent by the state and capital to accomplish this goal and to encourage a collective forgetting of the pandemic and the progressive possibilities it heralded that were discussed by Žižek (2020) and intimated by Dorling (2020) and Reisz (2020). However, the fear is that this apparent largesse by capital would be swiftly followed by a new form of austerity that mobilises the language of sacrifice to re-establish the economy in a form that mirrors its predecessor, with all its tensions, contradictions and inequalities—the re-formation of a post-neo-liberal economy. However, such an economy could be one in which the patterns of online working that were necessary during the pandemic and the resultant digital relations become part of the new normal. The irony is that whilst digitalisation and the use of IT is thought to lead to efficiency gains it encounters a 'productive paradox' whereby the reverse occurs. An example would be the increased workload generated during the pandemic for academic staff when British universities sought to place courses online. Rather than facilitating a more efficient labour process, these experiences led to its intensification. Such practices may well exacerbate the manner in which all of life becomes work and whereby the promise of more free-time evaporates.

Theodoropoulos (2020, online) notes that Agamben 'warns that the atomising, solidarity and community-destroying trends that existed before the pandemic will probably be exacerbated during and after it'. In addition, Agamben (2020, online) suggests that

> it is very likely that governments, after the end of the health emergency, will seek to continue the experiments that they have not yet managed to carry out: that universities and schools close and only give online lessons, that we stop gathering and talking for political or cultural reasons and only exchange digital messages, that as far as possible machines replace all contact—all contagion—between human beings.

The point is that social distancing and the marginalisation of our collective responsibilities existed prior to the pandemic—we only need think of gated communities. Whilst it may be an exaggeration to suggest that universities and schools will close and be replaced by virtual engagements there may well be a move in this direction with the forms of atomisation that surround such individualised labour.

Importantly, whatever the outcome and the manner in which the COVID-19 crisis is resolved it is as well to remember the adaptive capacities of capital. That is to say the facility of capital to adjust to the circumstances it faces as well as the role of the state and supranational organisations, such as the World Bank, IMF and so on, in securing the conditions favourable for the reproduction of capital. We need only bear in mind the manner in which capital morphed from a social democratic stance towards that of neo-liberalism in the late 1970s to illustrate its adaptability to changing circumstances. Importantly and to reiterate an argument made earlier in the book many of the issues discussed in this postscript cannot be resolved by education alone and necessitate the development of a non-reformist politics in pursuance of a socially just society. Nevertheless it is possible to intimate towards the parameters in which VET could be placed. Whilst 'gig' and project-based employment may not represent the working lives of the majority of the employed, it may presage the future. Boutang (2011: 60) noted that

> Marx, in his study of capital, did not study the largest working population in England at that time that comprised of domestic servants, but rather focused upon a much smaller group of factory workers that anticipated the emergent hegemony of industrial capitalism.

Following the pandemic flexible and adaptable labour may become an increasingly prominent feature of working lives, enabling such worker to engage with VET and education as a right becomes an aspect of distributive justice. This would be a VET that refuses a narrow instrumentalism and technicism, being one that can open up dialogue and considers the social obligations surrounding vocational practice. It would also take into account the relationship of occupations to wider society and the manner in which these could contribute to a socially just society. The point is not to provide a blueprint to be followed in the vein of policy science but rather to open up debate and raise questions—not least in the struggle for gentler, kinder and more egalitarian society.

References

Agamben, G. 2020. Normalising the State of Exception under the COVID-19 Epidemic. https://non.copyriot.com/giorgio-agamben-normalising-the-state-of-exception-under-the-covid-19-epidemic/

Allman, P., P. McLaren, and G. Rikowski. 2003. After the Box People. In *Yesterday's Dreams: International and Critical Perspectives on Education and Social Class*, ed. J. Freeman-Moir and A. Scott, 149–179. Christchurch: Canterbury University Press.

Berry, C. 2020. The COVID-19 Pandemic will Change Everything—for Better or Worse. https://www.versobooks.com/blogs/4613-the-covid-19-pandemic-will-change-everything-for-better-or-worse?utm_source=Master+List&utm_campaign=53bb38c6d5-UK+Direct%3A+Verso+blog+response&utm_medium=email&utm_term=0_1f96ba5fab-53bb38c6d5-410277665&mc_cid=53bb38c6d5&mc_eid=c622b1ec82

Boutang, Y.M. 2011. *Cognitive Capitalism*. Cambridge: Polity.

Butler, J. 2020. *Capitalism Has its Limits*. https://www.versobooks.com/blogs/4603-capitalism-has-its-limits?utm_source=Master+List&utm_campaign=53bb38c6d5-UK+Direct%3A+Verso+blog+response&utm_medium=email&utm_term=0_1f96ba5fab-53bb38c6d5-410277665&mc_cid=53bb38c6d5&mc_eid=c622b1ec82

The Care Collective. 2020. COVID-19 Pandemic: A Crisis of Care. https://www.versobooks.com/blogs/4617-covid-19-pandemic-a-crisis-of-care?utm_source=Master+List&utm_campaign=53bb38c6d5-UK+Direct%3A+Verso+blog+response&utm_medium=email&utm_term=0_1f96ba5fab-53bb38c6d5-410277665&mc_cid=53bb38c6d5&mc_eid=c622b1ec82

Cohen, P. 2020. Going Viral? Cartographies of Panic and Precaution In the Age of Precarity Phil Cohen. https://philcohenworks.com/going-viral-cartographies-of-panic-and-precaution-in-the-age-of-precarity/

Dorling, D. 2020. *Slowdown: The End of the Great Acceleration*. New Haven: Yale.

Fuller, A., and L. Unwin. 2003. Learning as Apprentices in the Contemporary UK Workplace: Creating and Managing Expansive and Restrictive Participation. *Journal of Education and Work* 16 (4): 407–426. https://doi.org/10.1080/1363908032000093012.

Galvin, R., and Healy, N. 2020. The Green New Deal Is More Relevant than Ever. https://blogs.scientificamerican.com/observations/the-green-new-deal-is-more-relevant-than-ever/

Gambuto, J. 2020. Prepare for the Ultimate Gaslighting. https://forge.medium.com/prepare-for-the-ultimate-gaslighting-6a8ce3f0a0e0

Gramsci, A. 1971. *Prison Notebooks*. London: Lawrence and Wishart.

The Green New Deal Group. 2008. *A Green New Deal*. London: The New Economics Foundation.

Harvey, D. 2014. *Seventeen Contradictions and the End of capitalism*. London: profile books.

Klein, N. 2007. *The Shock Doctrine*. London: Penguin.

Powell, D. 2019. Rebooting the Uk's Green New Deal. https://neweconomics.org/2019/01/a-green-new-deal-for-the-uk

Reisz, M. 2020. Danny Dorling: 'Slowdown' Brings end to 'Rampant Capitalism'. *Times Higher Education* 2, 454. April 16–19. https://www.timeshighereducation.com/people/danny-dorling-slowdown-brings-end-rampant-capitalism

Sierra Club. n.d.-a. Growing the Green New Deal. https://www.sierraclub.org/trade/what-green-new-deal

———. n.d.-b. People's Bailout. https://www.sierraclub.org/trade/peoples-bailout

Solis, M. 2020. Coronavirus Is the Perfect Disaster for 'Disaster Capitalism'. https://www.vice.com/en_uk/article/5dmqyk/naomi-klein-interview-on-coronavirus-and-disaster-capitalism-shock-doctrine

Theodoropoulos, P. 2020. Agamben, Zizek, and the Virus: An Attempt at an Anti-Authoritarian Synthesis. *Interregnum*. https://interregnum.live/2020/04/02/agamben-zizek-and-the-virus-an-attempt-at-an-anti-authoritarian-synthesis/

Tooze, A. 2020. Shockwave. *London Review of Books* 42(8). https://www.lrb.co.uk/the-paper/v42/n08/adam-tooze/shockwave

Žižek, S. 2020. Monitor and Punish? Yes Please. *Los Angeles Review of Books*. http://thephilosophicalsalon.com/monitor-and-punish-yes-please/?fbclid=IwAR1tBVPAAyZbAbjf1VrPDxQ-rqFxJdxmyanH4di8X-JEqsyDJode51ghL

Zuboff, S. 2019. *The Age of Surveillance Capitalism*. London: Profile Books.

Index

J. Avis, *Vocational Education in the Fourth Industrial Revolution*,
https://doi.org/10.1007/978-3-030-52032-8

The manufacturer's authorised representative in the EU is Springer Nature Customer Service Centre GmbH, Europaplatz 3, 69115 Heidelberg, Germany. If you have any concerns regarding our products, please contact ProductSafety@springernature.com

Printed and bound by CPI Group (UK) Ltd, Croydon, CR0 4YY
30/09/2026
02225711-0002